THE PRACTICAL SAFETY SERIES

THE RAILWAY CONTROL CENTRE COMMUNICATION PLAYBOOK

REMOTE OPERATIONAL CONTROL, PASSENGER ASSISTANCE AND DISTRESS COMMUNICATION

SAFETY-CRITICAL | ACTIVE LISTENING | DISTRESS

PASSENGER ASSIST | HANDOVER | WELFARE

LISTEN | LOCATE | CONFIRM | PROTECT | COORDINATE

ROBERT PORTER

HOW TO USE THIS PLAYBOOK

Read the principles, then use the paired practice pages and control cards for training, simulation, coaching and reflective review.

- **Use current role competence and local emergency procedures.**
- **Separate human support from formal operational authority.**
- **Practise handovers and system failure before live demand.**

NON-NEGOTIABLE

No phrase guarantees safety. Connection and communication must lead to location confirmation, risk control, competent local response and accepted handover.

SCOPE AND LIMITATIONS

This book covers common principles across operational, station, security, passenger-assistance and contact-centre functions. Authority, systems and language differ.

- **It does not authorise signalling, electrical, train movement or emergency action.**
- **It does not replace suicide-intervention, safeguarding or clinical training.**
- **Scripts never override live risk, caller information or controlled procedure.**

CONTROLLED BOUNDARY

Use the current Rule Book, safety management system, Accessible Travel Policy, control manual and emergency instructions applicable to the role.

A NOTE FROM ROBERT

Remote control asks you to create an accurate picture without standing in the scene. The quality of the next decision depends on how well you hear both the facts and the person.

- **Locate before assuming.**
- **Validate without inventing certainty.**
- **Transfer responsibility explicitly.**

ROBERT'S PERSPECTIVE

Empathy and operational discipline are not competing skills. The safest controllers know when each must lead and how to keep both present.

PART 1 | CONTROL ENVIRONMENTS

UNDERSTANDING THE CONTROL LANDSCAPE

Remote control functions connect different authorities, information systems and people who may never share the same view.

RAIL OPERATING CENTRES

Route control, signalling and operational functions use defined authority and formal communication to manage network risk..

- **Establish exact location, immediate danger, source and caller needs.**
- **Choose the communication mode, authorised response and explicit owner.**
- **Read back critical detail, record action and confirm safe handover.**

CONTROL PRINCIPLE

Use role-specific scripts as prompts, not substitutes for listening. Any emergency, safeguarding or safety-critical action must follow current controlled procedure.

RAIL OPERATING CENTRES - COMMUNICATION PRACTICE

Use this page for simulation, coaching or review. Do not enter real personal data in an uncontrolled workbook.

- **Caller, channel and exact location:** ______________________________
- **What is happening and immediate danger:** ________________________
- **Words that acknowledged and clarified:** __________________________
- **Operational action, authority and owner:** ________________________
- **Read-back, update and handover:** ________________________________
- **Welfare and learning:** ____

PRACTICE

A strong call record shows what was known, what remained uncertain, how risk was controlled and who explicitly accepted the next action.

TRAIN OPERATOR CONTROL

TOC control coordinates service, fleet, crew, stations, information and recovery within its own responsibilities..

- **Establish exact location, immediate danger, source and caller needs.**
- **Choose the communication mode, authorised response and explicit owner.**
- **Read back critical detail, record action and confirm safe handover.**

CONTROL PRINCIPLE

Use role-specific scripts as prompts, not substitutes for listening. Any emergency, safeguarding or safety-critical action must follow current controlled procedure.

TRAIN OPERATOR CONTROL - COMMUNICATION PRACTICE

Use this page for simulation, coaching or review. Do not enter real personal data in an uncontrolled workbook.

- **Caller, channel and exact location: ______________________________**
- **What is happening and immediate danger: ________________________**
- **Words that acknowledged and clarified: ___________________________**
- **Operational action, authority and owner: ________________________**
- **Read-back, update and handover: _______________________________**
- **Welfare and learning: ____**

PRACTICE

A strong call record shows what was known, what remained uncertain, how risk was controlled and who explicitly accepted the next action.

STATION AND SECURITY CONTROL

Local control integrates CCTV, assets, access, incidents and passenger movement without replacing specialist controllers..

- **Establish exact location, immediate danger, source and caller needs.**
- **Choose the communication mode, authorised response and explicit owner.**
- **Read back critical detail, record action and confirm safe handover.**

CONTROL PRINCIPLE

Use role-specific scripts as prompts, not substitutes for listening. Any emergency, safeguarding or safety-critical action must follow current controlled procedure.

STATION AND SECURITY CONTROL - COMMUNICATION PRACTICE

Use this page for simulation, coaching or review. Do not enter real personal data in an uncontrolled workbook.

- **Caller, channel and exact location: ______________________________**
- **What is happening and immediate danger: __________________________**
- **Words that acknowledged and clarified: ___________________________**
- **Operational action, authority and owner: _________________________**
- **Read-back, update and handover: _________________________________**
- **Welfare and learning: ____**

PRACTICE

A strong call record shows what was known, what remained uncertain, how risk was controlled and who explicitly accepted the next action.

PASSENGER ASSISTANCE CONTROL

Assistance centres arrange journeys and handovers; they do not authorise train movement or local operational access..

- **Establish exact location, immediate danger, source and caller needs.**
- **Choose the communication mode, authorised response and explicit owner.**
- **Read back critical detail, record action and confirm safe handover.**

CONTROL PRINCIPLE

Use role-specific scripts as prompts, not substitutes for listening. Any emergency, safeguarding or safety-critical action must follow current controlled procedure.

PASSENGER ASSISTANCE CONTROL - COMMUNICATION PRACTICE

Use this page for simulation, coaching or review. Do not enter real personal data in an uncontrolled workbook.

- **Caller, channel and exact location:** ______________________________
- **What is happening and immediate danger:** ________________________
- **Words that acknowledged and clarified:** __________________________
- **Operational action, authority and owner:** ________________________
- **Read-back, update and handover:** ________________________________
- **Welfare and learning:** ____

PRACTICE

A strong call record shows what was known, what remained uncertain, how risk was controlled and who explicitly accepted the next action.

CUSTOMER CONTACT CHANNELS

Telephone, help point, chat and social channels each change identity, location, privacy and emergency-response capability..

- **Establish exact location, immediate danger, source and caller needs.**
- **Choose the communication mode, authorised response and explicit owner.**
- **Read back critical detail, record action and confirm safe handover.**

CONTROL PRINCIPLE

Use role-specific scripts as prompts, not substitutes for listening. Any emergency, safeguarding or safety-critical action must follow current controlled procedure.

CUSTOMER CONTACT CHANNELS - COMMUNICATION PRACTICE

Use this page for simulation, coaching or review. Do not enter real personal data in an uncontrolled workbook.

- **Caller, channel and exact location:** ______________________________
- **What is happening and immediate danger:** ________________________
- **Words that acknowledged and clarified:** __________________________
- **Operational action, authority and owner:** ________________________
- **Read-back, update and handover:** _______________________________
- **Welfare and learning:** ____

PRACTICE

A strong call record shows what was known, what remained uncertain, how risk was controlled and who explicitly accepted the next action.

INTERFACE DISCIPLINE

Every centre must know who owns the decision, who supplies information and who confirms the handover..

- **Establish exact location, immediate danger, source and caller needs.**
- **Choose the communication mode, authorised response and explicit owner.**
- **Read back critical detail, record action and confirm safe handover.**

CONTROL PRINCIPLE

Use role-specific scripts as prompts, not substitutes for listening. Any emergency, safeguarding or safety-critical action must follow current controlled procedure.

INTERFACE DISCIPLINE - COMMUNICATION PRACTICE

Use this page for simulation, coaching or review. Do not enter real personal data in an uncontrolled workbook.

- **Caller, channel and exact location:** ______________________________
- **What is happening and immediate danger:** ________________________
- **Words that acknowledged and clarified:** __________________________
- **Operational action, authority and owner:** ________________________
- **Read-back, update and handover:** ________________________________
- **Welfare and learning:** ____

PRACTICE

A strong call record shows what was known, what remained uncertain, how risk was controlled and who explicitly accepted the next action.

PART 2 | REMOTE SITUATIONAL AWARENESS

BUILDING THE PICTURE THROUGH WORDS

The controller cannot assume that the caller sees, names or interprets the railway environment accurately.

LOCATION FIRST

Establish station, platform, direction, train, coach, access point and a recognisable landmark as early as the situation allows..

- **Establish exact location, immediate danger, source and caller needs.**
- **Choose the communication mode, authorised response and explicit owner.**
- **Read back critical detail, record action and confirm safe handover.**

CONTROL PRINCIPLE

Use role-specific scripts as prompts, not substitutes for listening. Any emergency, safeguarding or safety-critical action must follow current controlled procedure.

LOCATION FIRST - COMMUNICATION PRACTICE

Use this page for simulation, coaching or review. Do not enter real personal data in an uncontrolled workbook.

- **Caller, channel and exact location:** ______________________________
- **What is happening and immediate danger:** ________________________
- **Words that acknowledged and clarified:** __________________________
- **Operational action, authority and owner:** ________________________
- **Read-back, update and handover:** ________________________________
- **Welfare and learning:** ____

PRACTICE

A strong call record shows what was known, what remained uncertain, how risk was controlled and who explicitly accepted the next action.

WHAT IS HAPPENING NOW

Separate the current condition from how the event began or what the caller believes may happen..

- **Establish exact location, immediate danger, source and caller needs.**
- **Choose the communication mode, authorised response and explicit owner.**
- **Read back critical detail, record action and confirm safe handover.**

CONTROL PRINCIPLE

Use role-specific scripts as prompts, not substitutes for listening. Any emergency, safeguarding or safety-critical action must follow current controlled procedure.

WHAT IS HAPPENING NOW - COMMUNICATION PRACTICE

Use this page for simulation, coaching or review. Do not enter real personal data in an uncontrolled workbook.

- **Caller, channel and exact location:** ________________________________
- **What is happening and immediate danger:** ____________________________
- **Words that acknowledged and clarified:** _____________________________
- **Operational action, authority and owner:** ____________________________
- **Read-back, update and handover:** ___________________________________
- **Welfare and learning:** ____

PRACTICE

A strong call record shows what was known, what remained uncertain, how risk was controlled and who explicitly accepted the next action.

IMMEDIATE DANGER

Ask about train movement, track access, platform edge, electricity, fire, violence, injury and inability to move safely..

- **Establish exact location, immediate danger, source and caller needs.**
- **Choose the communication mode, authorised response and explicit owner.**
- **Read back critical detail, record action and confirm safe handover.**

CONTROL PRINCIPLE

Use role-specific scripts as prompts, not substitutes for listening. Any emergency, safeguarding or safety-critical action must follow current controlled procedure.

IMMEDIATE DANGER - COMMUNICATION PRACTICE

Use this page for simulation, coaching or review. Do not enter real personal data in an uncontrolled workbook.

- **Caller, channel and exact location:** ______________________________
- **What is happening and immediate danger:** ___________________________
- **Words that acknowledged and clarified:** ____________________________
- **Operational action, authority and owner:** __________________________
- **Read-back, update and handover:** __________________________________
- **Welfare and learning:** ____

PRACTICE

A strong call record shows what was known, what remained uncertain, how risk was controlled and who explicitly accepted the next action.

SOURCE AND CERTAINTY

Record what was directly seen, heard from another person, inferred or confirmed through systems and staff..

- **Establish exact location, immediate danger, source and caller needs.**
- **Choose the communication mode, authorised response and explicit owner.**
- **Read back critical detail, record action and confirm safe handover.**

CONTROL PRINCIPLE

Use role-specific scripts as prompts, not substitutes for listening. Any emergency, safeguarding or safety-critical action must follow current controlled procedure.

SOURCE AND CERTAINTY - COMMUNICATION PRACTICE

Use this page for simulation, coaching or review. Do not enter real personal data in an uncontrolled workbook.

- **Caller, channel and exact location:** ______________________________
- **What is happening and immediate danger:** ________________________
- **Words that acknowledged and clarified:** __________________________
- **Operational action, authority and owner:** ________________________
- **Read-back, update and handover:** _______________________________
- **Welfare and learning:** ____

PRACTICE

A strong call record shows what was known, what remained uncertain, how risk was controlled and who explicitly accepted the next action.

BACKGROUND INFORMATION

Listen for announcements, alarms, trains, wind, crowds, breathing and other cues without treating them as proof..

- **Establish exact location, immediate danger, source and caller needs.**
- **Choose the communication mode, authorised response and explicit owner.**
- **Read back critical detail, record action and confirm safe handover.**

CONTROL PRINCIPLE

Use role-specific scripts as prompts, not substitutes for listening. Any emergency, safeguarding or safety-critical action must follow current controlled procedure.

BACKGROUND INFORMATION - COMMUNICATION PRACTICE

Use this page for simulation, coaching or review. Do not enter real personal data in an uncontrolled workbook.

- **Caller, channel and exact location: ________________________________**
- **What is happening and immediate danger: ____________________________**
- **Words that acknowledged and clarified: ____________________________**
- **Operational action, authority and owner: __________________________**
- **Read-back, update and handover: __________________________________**
- **Welfare and learning: ____**

PRACTICE

A strong call record shows what was known, what remained uncertain, how risk was controlled and who explicitly accepted the next action.

UPDATING THE PICTURE

Reconfirm location, people and hazards whenever the call transfers, conditions change or new responders arrive..

- **Establish exact location, immediate danger, source and caller needs.**
- **Choose the communication mode, authorised response and explicit owner.**
- **Read back critical detail, record action and confirm safe handover.**

CONTROL PRINCIPLE

Use role-specific scripts as prompts, not substitutes for listening. Any emergency, safeguarding or safety-critical action must follow current controlled procedure.

UPDATING THE PICTURE - COMMUNICATION PRACTICE

Use this page for simulation, coaching or review. Do not enter real personal data in an uncontrolled workbook.

- **Caller, channel and exact location:** ______________________
- **What is happening and immediate danger:** ______________________
- **Words that acknowledged and clarified:** ______________________
- **Operational action, authority and owner:** ______________________
- **Read-back, update and handover:** ______________________
- **Welfare and learning:** ______________________

PRACTICE

A strong call record shows what was known, what remained uncertain, how risk was controlled and who explicitly accepted the next action.

PART 3 | FIRST CONTACT

THE FIRST SECONDS OF THE CALL

Identity, location, danger and callback establish the minimum connection needed to coordinate a response.

INTRODUCE THE FUNCTION

Use a calm identity and purpose so the person knows whom they reached and what the centre can do..

- **Establish exact location, immediate danger, source and caller needs.**
- **Choose the communication mode, authorised response and explicit owner.**
- **Read back critical detail, record action and confirm safe handover.**

CONTROL PRINCIPLE

Use role-specific scripts as prompts, not substitutes for listening. Any emergency, safeguarding or safety-critical action must follow current controlled procedure.

INTRODUCE THE FUNCTION - COMMUNICATION PRACTICE

Use this page for simulation, coaching or review. Do not enter real personal data in an uncontrolled workbook.

- **Caller, channel and exact location: ______________________________**
- **What is happening and immediate danger: ________________________**
- **Words that acknowledged and clarified: __________________________**
- **Operational action, authority and owner: ________________________**
- **Read-back, update and handover: _______________________________**
- **Welfare and learning: ______________________________________**

PRACTICE

A strong call record shows what was known, what remained uncertain, how risk was controlled and who explicitly accepted the next action.

ASK EXACT LOCATION

Use one short question, then clarify with platform, sign, entrance, coach or help-point number..

- **Establish exact location, immediate danger, source and caller needs.**
- **Choose the communication mode, authorised response and explicit owner.**
- **Read back critical detail, record action and confirm safe handover.**

CONTROL PRINCIPLE

Use role-specific scripts as prompts, not substitutes for listening. Any emergency, safeguarding or safety-critical action must follow current controlled procedure.

ASK EXACT LOCATION - COMMUNICATION PRACTICE

Use this page for simulation, coaching or review. Do not enter real personal data in an uncontrolled workbook.

- **Caller, channel and exact location:** ______________________________
- **What is happening and immediate danger:** ________________________
- **Words that acknowledged and clarified:** __________________________
- **Operational action, authority and owner:** ________________________
- **Read-back, update and handover:** _______________________________
- **Welfare and learning:** ____

PRACTICE

A strong call record shows what was known, what remained uncertain, how risk was controlled and who explicitly accepted the next action.

CHECK IMMEDIATE DANGER

Ask directly whether the caller or anyone else is at immediate risk before taking a full history..

- **Establish exact location, immediate danger, source and caller needs.**
- **Choose the communication mode, authorised response and explicit owner.**
- **Read back critical detail, record action and confirm safe handover.**

CONTROL PRINCIPLE

Use role-specific scripts as prompts, not substitutes for listening. Any emergency, safeguarding or safety-critical action must follow current controlled procedure.

CHECK IMMEDIATE DANGER - COMMUNICATION PRACTICE

Use this page for simulation, coaching or review. Do not enter real personal data in an uncontrolled workbook.

- **Caller, channel and exact location:** ________________________________
- **What is happening and immediate danger:** __________________________
- **Words that acknowledged and clarified:** ____________________________
- **Operational action, authority and owner:** __________________________
- **Read-back, update and handover:** __________________________________
- **Welfare and learning:** ____

PRACTICE

A strong call record shows what was known, what remained uncertain, how risk was controlled and who explicitly accepted the next action.

ESTABLISH CALLBACK

Confirm the number or fallback channel where systems and policy permit; explain if the call is recorded..

- **Establish exact location, immediate danger, source and caller needs.**
- **Choose the communication mode, authorised response and explicit owner.**
- **Read back critical detail, record action and confirm safe handover.**

CONTROL PRINCIPLE

Use role-specific scripts as prompts, not substitutes for listening. Any emergency, safeguarding or safety-critical action must follow current controlled procedure.

ESTABLISH CALLBACK - COMMUNICATION PRACTICE

Use this page for simulation, coaching or review. Do not enter real personal data in an uncontrolled workbook.

- **Caller, channel and exact location:** ______________________________
- **What is happening and immediate danger:** ______________________________
- **Words that acknowledged and clarified:** ______________________________
- **Operational action, authority and owner:** ______________________________
- **Read-back, update and handover:** ______________________________
- **Welfare and learning:** ______________________________

PRACTICE

A strong call record shows what was known, what remained uncertain, how risk was controlled and who explicitly accepted the next action.

SET EXPECTATIONS

Say that you will ask short questions and arrange the correct help rather than promising a specific outcome..

- **Establish exact location, immediate danger, source and caller needs.**
- **Choose the communication mode, authorised response and explicit owner.**
- **Read back critical detail, record action and confirm safe handover.**

CONTROL PRINCIPLE

Use role-specific scripts as prompts, not substitutes for listening. Any emergency, safeguarding or safety-critical action must follow current controlled procedure.

SET EXPECTATIONS - COMMUNICATION PRACTICE

Use this page for simulation, coaching or review. Do not enter real personal data in an uncontrolled workbook.

- **Caller, channel and exact location:** ______________________________
- **What is happening and immediate danger:** ________________________
- **Words that acknowledged and clarified:** __________________________
- **Operational action, authority and owner:** ________________________
- **Read-back, update and handover:** ________________________________
- **Welfare and learning:** ____

PRACTICE

A strong call record shows what was known, what remained uncertain, how risk was controlled and who explicitly accepted the next action.

CONTROL WITHOUT DOMINATION

Interrupt only for immediate safety or critical clarification; otherwise let the person complete the essential account..

- **Establish exact location, immediate danger, source and caller needs.**
- **Choose the communication mode, authorised response and explicit owner.**
- **Read back critical detail, record action and confirm safe handover.**

CONTROL PRINCIPLE

Use role-specific scripts as prompts, not substitutes for listening. Any emergency, safeguarding or safety-critical action must follow current controlled procedure.

CONTROL WITHOUT DOMINATION - COMMUNICATION PRACTICE

Use this page for simulation, coaching or review. Do not enter real personal data in an uncontrolled workbook.

- **Caller, channel and exact location:** ______________________________
- **What is happening and immediate danger:** ________________________
- **Words that acknowledged and clarified:** __________________________
- **Operational action, authority and owner:** ________________________
- **Read-back, update and handover:** _________________________________
- **Welfare and learning:** ____

PRACTICE

A strong call record shows what was known, what remained uncertain, how risk was controlled and who explicitly accepted the next action.

HEARING FACT, MEANING AND EMOTION

Listening is operational work: it reveals location, risk, misunderstanding, distress and the information the caller cannot yet organise.

GIVE SPACE

Allow the caller to finish a coherent point and use silence without abandoning the connection..

- **Establish exact location, immediate danger, source and caller needs.**
- **Choose the communication mode, authorised response and explicit owner.**
- **Read back critical detail, record action and confirm safe handover.**

CONTROL PRINCIPLE

Use role-specific scripts as prompts, not substitutes for listening. Any emergency, safeguarding or safety-critical action must follow current controlled procedure.

GIVE SPACE - COMMUNICATION PRACTICE

Use this page for simulation, coaching or review. Do not enter real personal data in an uncontrolled workbook.

- **Caller, channel and exact location:** ______________________________
- **What is happening and immediate danger:** ________________________
- **Words that acknowledged and clarified:** __________________________
- **Operational action, authority and owner:** ________________________
- **Read-back, update and handover:** _______________________________
- **Welfare and learning:** ____

PRACTICE

A strong call record shows what was known, what remained uncertain, how risk was controlled and who explicitly accepted the next action.

REFLECT KEY WORDS

Repeat a significant phrase in neutral language so the person can confirm or correct your understanding..

- **Establish exact location, immediate danger, source and caller needs.**
- **Choose the communication mode, authorised response and explicit owner.**
- **Read back critical detail, record action and confirm safe handover.**

CONTROL PRINCIPLE

Use role-specific scripts as prompts, not substitutes for listening. Any emergency, safeguarding or safety-critical action must follow current controlled procedure.

REFLECT KEY WORDS - COMMUNICATION PRACTICE

Use this page for simulation, coaching or review. Do not enter real personal data in an uncontrolled workbook.

- **Caller, channel and exact location:** ______________________________
- **What is happening and immediate danger:** ________________________
- **Words that acknowledged and clarified:** __________________________
- **Operational action, authority and owner:** ________________________
- **Read-back, update and handover:** ________________________________
- **Welfare and learning:** ____

PRACTICE

A strong call record shows what was known, what remained uncertain, how risk was controlled and who explicitly accepted the next action.

SUMMARISE

Bring a long account into location, event, risk, people and requested help without deleting uncertainty..

- **Establish exact location, immediate danger, source and caller needs.**
- **Choose the communication mode, authorised response and explicit owner.**
- **Read back critical detail, record action and confirm safe handover.**

CONTROL PRINCIPLE

Use role-specific scripts as prompts, not substitutes for listening. Any emergency, safeguarding or safety-critical action must follow current controlled procedure.

SUMMARISE - COMMUNICATION PRACTICE

Use this page for simulation, coaching or review. Do not enter real personal data in an uncontrolled workbook.

- **Caller, channel and exact location:** ______________________________
- **What is happening and immediate danger:** ______________________________
- **Words that acknowledged and clarified:** ______________________________
- **Operational action, authority and owner:** ______________________________
- **Read-back, update and handover:** ______________________________
- **Welfare and learning:** ______________________________

PRACTICE

A strong call record shows what was known, what remained uncertain, how risk was controlled and who explicitly accepted the next action.

CLARIFY AMBIGUITY

Ask one specific question at a time and avoid stacking several demands into one sentence..

- **Establish exact location, immediate danger, source and caller needs.**
- **Choose the communication mode, authorised response and explicit owner.**
- **Read back critical detail, record action and confirm safe handover.**

CONTROL PRINCIPLE

Use role-specific scripts as prompts, not substitutes for listening. Any emergency, safeguarding or safety-critical action must follow current controlled procedure.

CLARIFY AMBIGUITY - COMMUNICATION PRACTICE

Use this page for simulation, coaching or review. Do not enter real personal data in an uncontrolled workbook.

- **Caller, channel and exact location:** ______________________________
- **What is happening and immediate danger:** ________________________
- **Words that acknowledged and clarified:** __________________________
- **Operational action, authority and owner:** ________________________
- **Read-back, update and handover:** _______________________________
- **Welfare and learning:** ____

PRACTICE

A strong call record shows what was known, what remained uncertain, how risk was controlled and who explicitly accepted the next action.

LISTEN FOR CHANGE

Notice speech, breathing, background, coherence or silence changing and reassess risk..

- **Establish exact location, immediate danger, source and caller needs.**
- **Choose the communication mode, authorised response and explicit owner.**
- **Read back critical detail, record action and confirm safe handover.**

CONTROL PRINCIPLE

Use role-specific scripts as prompts, not substitutes for listening. Any emergency, safeguarding or safety-critical action must follow current controlled procedure.

LISTEN FOR CHANGE - COMMUNICATION PRACTICE

Use this page for simulation, coaching or review. Do not enter real personal data in an uncontrolled workbook.

- **Caller, channel and exact location:** ______________________________
- **What is happening and immediate danger:** ______________________________
- **Words that acknowledged and clarified:** ______________________________
- **Operational action, authority and owner:** ______________________________
- **Read-back, update and handover:** ______________________________
- **Welfare and learning:** ______________________________

PRACTICE

A strong call record shows what was known, what remained uncertain, how risk was controlled and who explicitly accepted the next action.

AVOID PREMATURE SOLUTION

Do not offer a standard remedy before the person has described the immediate problem and required outcome..

- **Establish exact location, immediate danger, source and caller needs.**
- **Choose the communication mode, authorised response and explicit owner.**
- **Read back critical detail, record action and confirm safe handover.**

CONTROL PRINCIPLE

Use role-specific scripts as prompts, not substitutes for listening. Any emergency, safeguarding or safety-critical action must follow current controlled procedure.

AVOID PREMATURE SOLUTION - COMMUNICATION PRACTICE

Use this page for simulation, coaching or review. Do not enter real personal data in an uncontrolled workbook.

- **Caller, channel and exact location: ________________________________**
- **What is happening and immediate danger: ____________________________**
- **Words that acknowledged and clarified: ______________________________**
- **Operational action, authority and owner: ____________________________**
- **Read-back, update and handover: _________________________________**
- **Welfare and learning: ____**

PRACTICE

A strong call record shows what was known, what remained uncertain, how risk was controlled and who explicitly accepted the next action.

PART 5 | VALIDATION

ACKNOWLEDGE WITHOUT INVENTING FACT

Validation recognises experience and emotion; it does not confirm an uninvestigated allegation or promise that everything will be fine.

NAME THE IMPACT

Say that the event sounds frightening, upsetting or frustrating where that matches what the person expressed..

- **Establish exact location, immediate danger, source and caller needs.**
- **Choose the communication mode, authorised response and explicit owner.**
- **Read back critical detail, record action and confirm safe handover.**

CONTROL PRINCIPLE

Use role-specific scripts as prompts, not substitutes for listening. Any emergency, safeguarding or safety-critical action must follow current controlled procedure.

NAME THE IMPACT - COMMUNICATION PRACTICE

Use this page for simulation, coaching or review. Do not enter real personal data in an uncontrolled workbook.

- **Caller, channel and exact location:** ____________________
- **What is happening and immediate danger:** ____________________
- **Words that acknowledged and clarified:** ____________________
- **Operational action, authority and owner:** ____________________
- **Read-back, update and handover:** ____________________
- **Welfare and learning:** ____________________

PRACTICE

A strong call record shows what was known, what remained uncertain, how risk was controlled and who explicitly accepted the next action.

TAKE THE REPORT SERIOUSLY

Thank the person for reporting and explain the immediate action you can take..

- **Establish exact location, immediate danger, source and caller needs.**
- **Choose the communication mode, authorised response and explicit owner.**
- **Read back critical detail, record action and confirm safe handover.**

CONTROL PRINCIPLE

Use role-specific scripts as prompts, not substitutes for listening. Any emergency, safeguarding or safety-critical action must follow current controlled procedure.

TAKE THE REPORT SERIOUSLY - COMMUNICATION PRACTICE

Use this page for simulation, coaching or review. Do not enter real personal data in an uncontrolled workbook.

- **Caller, channel and exact location: ______________________________**
- **What is happening and immediate danger: ________________________**
- **Words that acknowledged and clarified: __________________________**
- **Operational action, authority and owner: ________________________**
- **Read-back, update and handover: _______________________________**
- **Welfare and learning: ____**

PRACTICE

A strong call record shows what was known, what remained uncertain, how risk was controlled and who explicitly accepted the next action.

VALIDATE AND VERIFY

Acknowledge their experience while separately checking operational facts, records and other accounts..

- **Establish exact location, immediate danger, source and caller needs.**
- **Choose the communication mode, authorised response and explicit owner.**
- **Read back critical detail, record action and confirm safe handover.**

CONTROL PRINCIPLE

Use role-specific scripts as prompts, not substitutes for listening. Any emergency, safeguarding or safety-critical action must follow current controlled procedure.

VALIDATE AND VERIFY - COMMUNICATION PRACTICE

Use this page for simulation, coaching or review. Do not enter real personal data in an uncontrolled workbook.

- **Caller, channel and exact location: ______________________________**
- **What is happening and immediate danger: ________________________**
- **Words that acknowledged and clarified: __________________________**
- **Operational action, authority and owner: ________________________**
- **Read-back, update and handover: ________________________________**
- **Welfare and learning: ____**

PRACTICE

A strong call record shows what was known, what remained uncertain, how risk was controlled and who explicitly accepted the next action.

OWN THE NEXT STEP

State what you will do, who must accept action and when the person should expect an update..

- **Establish exact location, immediate danger, source and caller needs.**
- **Choose the communication mode, authorised response and explicit owner.**
- **Read back critical detail, record action and confirm safe handover.**

CONTROL PRINCIPLE

Use role-specific scripts as prompts, not substitutes for listening. Any emergency, safeguarding or safety-critical action must follow current controlled procedure.

OWN THE NEXT STEP - COMMUNICATION PRACTICE

Use this page for simulation, coaching or review. Do not enter real personal data in an uncontrolled workbook.

- **Caller, channel and exact location:** ________________________________
- **What is happening and immediate danger:** __________________________
- **Words that acknowledged and clarified:** ____________________________
- **Operational action, authority and owner:** __________________________
- **Read-back, update and handover:** __________________________________
- **Welfare and learning:** ____

PRACTICE

A strong call record shows what was known, what remained uncertain, how risk was controlled and who explicitly accepted the next action.

AVOID FALSE REASSURANCE

Do not say there is no danger or that help will arrive by a fixed time unless confirmed..

- **Establish exact location, immediate danger, source and caller needs.**
- **Choose the communication mode, authorised response and explicit owner.**
- **Read back critical detail, record action and confirm safe handover.**

CONTROL PRINCIPLE

Use role-specific scripts as prompts, not substitutes for listening. Any emergency, safeguarding or safety-critical action must follow current controlled procedure.

AVOID FALSE REASSURANCE - COMMUNICATION PRACTICE

Use this page for simulation, coaching or review. Do not enter real personal data in an uncontrolled workbook.

- **Caller, channel and exact location:** ______________________________
- **What is happening and immediate danger:** ___________________________
- **Words that acknowledged and clarified:** ____________________________
- **Operational action, authority and owner:** ___________________________
- **Read-back, update and handover:** ________________________________
- **Welfare and learning:** ____

PRACTICE

A strong call record shows what was known, what remained uncertain, how risk was controlled and who explicitly accepted the next action.

AVOID DISMISSIVE LANGUAGE

Phrases such as calm down, do not worry or at least nobody was hurt can increase isolation and anger..

- **Establish exact location, immediate danger, source and caller needs.**
- **Choose the communication mode, authorised response and explicit owner.**
- **Read back critical detail, record action and confirm safe handover.**

CONTROL PRINCIPLE

Use role-specific scripts as prompts, not substitutes for listening. Any emergency, safeguarding or safety-critical action must follow current controlled procedure.

AVOID DISMISSIVE LANGUAGE - COMMUNICATION PRACTICE

Use this page for simulation, coaching or review. Do not enter real personal data in an uncontrolled workbook.

- **Caller, channel and exact location: ______________________________**
- **What is happening and immediate danger: ______________________________**
- **Words that acknowledged and clarified: ______________________________**
- **Operational action, authority and owner: ______________________________**
- **Read-back, update and handover: ______________________________**
- **Welfare and learning: ______________________________**

PRACTICE

A strong call record shows what was known, what remained uncertain, how risk was controlled and who explicitly accepted the next action.

PART 6 | DISTRESS COMMUNICATION

CALM CONNECTION, CLEAR SAFETY ACTION

A distressed person may need emotional containment before they can provide ordered information or follow a complex instruction.

SLOW YOUR OWN DELIVERY

Use a lower pace, short sentences and one question at a time without speaking as if the person is a child..

- **Establish exact location, immediate danger, source and caller needs.**
- **Choose the communication mode, authorised response and explicit owner.**
- **Read back critical detail, record action and confirm safe handover.**

CONTROL PRINCIPLE

Use role-specific scripts as prompts, not substitutes for listening. Any emergency, safeguarding or safety-critical action must follow current controlled procedure.

SLOW YOUR OWN DELIVERY - COMMUNICATION PRACTICE

Use this page for simulation, coaching or review. Do not enter real personal data in an uncontrolled workbook.

- **Caller, channel and exact location:** ______________________
- **What is happening and immediate danger:** ____________________
- **Words that acknowledged and clarified:** ____________________
- **Operational action, authority and owner:** __________________
- **Read-back, update and handover:** ________________________
- **Welfare and learning:** ______________________________

PRACTICE

A strong call record shows what was known, what remained uncertain, how risk was controlled and who explicitly accepted the next action.

USE THE PERSON'S NAME

Use it naturally where known and consented, without repeating it mechanically..

- **Establish exact location, immediate danger, source and caller needs.**
- **Choose the communication mode, authorised response and explicit owner.**
- **Read back critical detail, record action and confirm safe handover.**

CONTROL PRINCIPLE

Use role-specific scripts as prompts, not substitutes for listening. Any emergency, safeguarding or safety-critical action must follow current controlled procedure.

USE THE PERSON'S NAME - COMMUNICATION PRACTICE

Use this page for simulation, coaching or review. Do not enter real personal data in an uncontrolled workbook.

- **Caller, channel and exact location: ________________________________**
- **What is happening and immediate danger: _________________________**
- **Words that acknowledged and clarified: ____________________________**
- **Operational action, authority and owner: __________________________**
- **Read-back, update and handover: ___________________________________**
- **Welfare and learning: ____**

PRACTICE

A strong call record shows what was known, what remained uncertain, how risk was controlled and who explicitly accepted the next action.

ORIENT TO THE PRESENT

Ask what they can see, where they are and what danger exists now rather than demanding the whole history..

- **Establish exact location, immediate danger, source and caller needs.**
- **Choose the communication mode, authorised response and explicit owner.**
- **Read back critical detail, record action and confirm safe handover.**

CONTROL PRINCIPLE

Use role-specific scripts as prompts, not substitutes for listening. Any emergency, safeguarding or safety-critical action must follow current controlled procedure.

ORIENT TO THE PRESENT - COMMUNICATION PRACTICE

Use this page for simulation, coaching or review. Do not enter real personal data in an uncontrolled workbook.

- **Caller, channel and exact location:** ________________________________
- **What is happening and immediate danger:** ____________________________
- **Words that acknowledged and clarified:** ____________________________
- **Operational action, authority and owner:** ____________________________
- **Read-back, update and handover:** ________________________________
- **Welfare and learning:** ______________________________________

PRACTICE

A strong call record shows what was known, what remained uncertain, how risk was controlled and who explicitly accepted the next action.

OFFER ONE SAFE STEP

Suggest a small action only when its route and destination are known to be safe..

- **Establish exact location, immediate danger, source and caller needs.**
- **Choose the communication mode, authorised response and explicit owner.**
- **Read back critical detail, record action and confirm safe handover.**

CONTROL PRINCIPLE

Use role-specific scripts as prompts, not substitutes for listening. Any emergency, safeguarding or safety-critical action must follow current controlled procedure.

OFFER ONE SAFE STEP - COMMUNICATION PRACTICE

Use this page for simulation, coaching or review. Do not enter real personal data in an uncontrolled workbook.

- **Caller, channel and exact location: ______________________________**
- **What is happening and immediate danger: ________________________**
- **Words that acknowledged and clarified: __________________________**
- **Operational action, authority and owner: ________________________**
- **Read-back, update and handover: ________________________________**
- **Welfare and learning: ____**

PRACTICE

A strong call record shows what was known, what remained uncertain, how risk was controlled and who explicitly accepted the next action.

MAINTAIN CONNECTION

Explain pauses, continue checking response and arrange help without leaving unexplained silence..

- **Establish exact location, immediate danger, source and caller needs.**
- **Choose the communication mode, authorised response and explicit owner.**
- **Read back critical detail, record action and confirm safe handover.**

CONTROL PRINCIPLE

Use role-specific scripts as prompts, not substitutes for listening. Any emergency, safeguarding or safety-critical action must follow current controlled procedure.

MAINTAIN CONNECTION - COMMUNICATION PRACTICE

Use this page for simulation, coaching or review. Do not enter real personal data in an uncontrolled workbook.

- **Caller, channel and exact location:** ______________________________
- **What is happening and immediate danger:** ________________________
- **Words that acknowledged and clarified:** __________________________
- **Operational action, authority and owner:** ________________________
- **Read-back, update and handover:** ________________________________
- **Welfare and learning:** ____

> **PRACTICE**
>
> **A strong call record shows what was known, what remained uncertain, how risk was controlled and who explicitly accepted the next action.**

TRANSFER WITH CARE

Tell the person who is joining, why, what has been shared and whether you will remain connected..

- **Establish exact location, immediate danger, source and caller needs.**
- **Choose the communication mode, authorised response and explicit owner.**
- **Read back critical detail, record action and confirm safe handover.**

CONTROL PRINCIPLE

Use role-specific scripts as prompts, not substitutes for listening. Any emergency, safeguarding or safety-critical action must follow current controlled procedure.

TRANSFER WITH CARE - COMMUNICATION PRACTICE

Use this page for simulation, coaching or review. Do not enter real personal data in an uncontrolled workbook.

- **Caller, channel and exact location:** ______________________________
- **What is happening and immediate danger:** ________________________
- **Words that acknowledged and clarified:** __________________________
- **Operational action, authority and owner:** ________________________
- **Read-back, update and handover:** ________________________________
- **Welfare and learning:** ____

PRACTICE

A strong call record shows what was known, what remained uncertain, how risk was controlled and who explicitly accepted the next action.

PART 7 | SUICIDE AND SELF-HARM

ASK DIRECTLY, LISTEN, ESCALATE

Direct, compassionate questions can open disclosure; the controller must then maintain connection and activate the trained emergency process.

RECOGNISE CONCERN

Statements of hopelessness, farewell language, dangerous position or disclosed intent require immediate trained assessment..

- **Establish exact location, immediate danger, source and caller needs.**
- **Choose the communication mode, authorised response and explicit owner.**
- **Read back critical detail, record action and confirm safe handover.**

CONTROL PRINCIPLE

Use role-specific scripts as prompts, not substitutes for listening. Any emergency, safeguarding or safety-critical action must follow current controlled procedure.

RECOGNISE CONCERN - COMMUNICATION PRACTICE

Use this page for simulation, coaching or review. Do not enter real personal data in an uncontrolled workbook.

- **Caller, channel and exact location:** ______________________________
- **What is happening and immediate danger:** ________________________
- **Words that acknowledged and clarified:** __________________________
- **Operational action, authority and owner:** ________________________
- **Read-back, update and handover:** ________________________________
- **Welfare and learning:** ____

PRACTICE

A strong call record shows what was known, what remained uncertain, how risk was controlled and who explicitly accepted the next action.

ASK DIRECTLY

Where concerned, ask clearly whether the person is thinking about suicide or harming themselves; do not hide behind vague euphemism..

- **Establish exact location, immediate danger, source and caller needs.**
- **Choose the communication mode, authorised response and explicit owner.**
- **Read back critical detail, record action and confirm safe handover.**

CONTROL PRINCIPLE

Use role-specific scripts as prompts, not substitutes for listening. Any emergency, safeguarding or safety-critical action must follow current controlled procedure.

ASK DIRECTLY - COMMUNICATION PRACTICE

Use this page for simulation, coaching or review. Do not enter real personal data in an uncontrolled workbook.

- **Caller, channel and exact location:** ______________________________
- **What is happening and immediate danger:** ________________________
- **Words that acknowledged and clarified:** __________________________
- **Operational action, authority and owner:** ________________________
- **Read-back, update and handover:** _______________________________
- **Welfare and learning:** ____

> **PRACTICE**
>
> **A strong call record shows what was known, what remained uncertain, how risk was controlled and who explicitly accepted the next action.**

ESTABLISH IMMEDIACY

Ask about current location, immediate intention, dangerous access, whether they are alone and whether action has begun..

- **Establish exact location, immediate danger, source and caller needs.**
- **Choose the communication mode, authorised response and explicit owner.**
- **Read back critical detail, record action and confirm safe handover.**

CONTROL PRINCIPLE

Use role-specific scripts as prompts, not substitutes for listening. Any emergency, safeguarding or safety-critical action must follow current controlled procedure.

ESTABLISH IMMEDIACY - COMMUNICATION PRACTICE

Use this page for simulation, coaching or review. Do not enter real personal data in an uncontrolled workbook.

- **Caller, channel and exact location:** ______________________________
- **What is happening and immediate danger:** ______________________________
- **Words that acknowledged and clarified:** ______________________________
- **Operational action, authority and owner:** ______________________________
- **Read-back, update and handover:** ______________________________
- **Welfare and learning:** ______________________________

PRACTICE

A strong call record shows what was known, what remained uncertain, how risk was controlled and who explicitly accepted the next action.

LISTEN WITHOUT JUDGEMENT

Do not debate, shame, guilt or rush into simplistic solutions; acknowledge the pain expressed..

- **Establish exact location, immediate danger, source and caller needs.**
- **Choose the communication mode, authorised response and explicit owner.**
- **Read back critical detail, record action and confirm safe handover.**

CONTROL PRINCIPLE

Use role-specific scripts as prompts, not substitutes for listening. Any emergency, safeguarding or safety-critical action must follow current controlled procedure.

LISTEN WITHOUT JUDGEMENT - COMMUNICATION PRACTICE

Use this page for simulation, coaching or review. Do not enter real personal data in an uncontrolled workbook.

- **Caller, channel and exact location:** ______________________________
- **What is happening and immediate danger:** ___________________________
- **Words that acknowledged and clarified:** ____________________________
- **Operational action, authority and owner:** ___________________________
- **Read-back, update and handover:** ________________________________
- **Welfare and learning:** ____

PRACTICE

A strong call record shows what was known, what remained uncertain, how risk was controlled and who explicitly accepted the next action.

DO NOT PROMISE SECRECY

Explain that information may need sharing to protect life and involve emergency or local responders..

- **Establish exact location, immediate danger, source and caller needs.**
- **Choose the communication mode, authorised response and explicit owner.**
- **Read back critical detail, record action and confirm safe handover.**

CONTROL PRINCIPLE

Use role-specific scripts as prompts, not substitutes for listening. Any emergency, safeguarding or safety-critical action must follow current controlled procedure.

DO NOT PROMISE SECRECY - COMMUNICATION PRACTICE

Use this page for simulation, coaching or review. Do not enter real personal data in an uncontrolled workbook.

- **Caller, channel and exact location:** ______________________________
- **What is happening and immediate danger:** ________________________
- **Words that acknowledged and clarified:** ___________________________
- **Operational action, authority and owner:** ________________________
- **Read-back, update and handover:** _______________________________
- **Welfare and learning:** ____

PRACTICE

A strong call record shows what was known, what remained uncertain, how risk was controlled and who explicitly accepted the next action.

SUSTAIN AND HAND OVER

Keep connection where safe and required, update responders and transfer only when responsibility is clearly accepted..

- **Establish exact location, immediate danger, source and caller needs.**
- **Choose the communication mode, authorised response and explicit owner.**
- **Read back critical detail, record action and confirm safe handover.**

CONTROL PRINCIPLE

Use role-specific scripts as prompts, not substitutes for listening. Any emergency, safeguarding or safety-critical action must follow current controlled procedure.

SUSTAIN AND HAND OVER - COMMUNICATION PRACTICE

Use this page for simulation, coaching or review. Do not enter real personal data in an uncontrolled workbook.

- **Caller, channel and exact location: ______________________________**
- **What is happening and immediate danger: ______________________________**
- **Words that acknowledged and clarified: ______________________________**
- **Operational action, authority and owner: ______________________________**
- **Read-back, update and handover: ______________________________**
- **Welfare and learning: ______________________________**

PRACTICE

A strong call record shows what was known, what remained uncertain, how risk was controlled and who explicitly accepted the next action.

PART 8 | SAFETY-CRITICAL COMMUNICATION

ACCURATE, BRIEF AND CLEAR

Operational safety messages require standard terminology, disciplined transmission and closed-loop confirmation.

IDENTITY AND AUTHORITY

Confirm who is speaking, role, location and authority before relying on an instruction or report..

- **Establish exact location, immediate danger, source and caller needs.**
- **Choose the communication mode, authorised response and explicit owner.**
- **Read back critical detail, record action and confirm safe handover.**

CONTROL PRINCIPLE

Use role-specific scripts as prompts, not substitutes for listening. Any emergency, safeguarding or safety-critical action must follow current controlled procedure.

IDENTITY AND AUTHORITY - COMMUNICATION PRACTICE

Use this page for simulation, coaching or review. Do not enter real personal data in an uncontrolled workbook.

- **Caller, channel and exact location:** ______________________________
- **What is happening and immediate danger:** ________________________
- **Words that acknowledged and clarified:** __________________________
- **Operational action, authority and owner:** ________________________
- **Read-back, update and handover:** _______________________________
- **Welfare and learning:** ____

PRACTICE

A strong call record shows what was known, what remained uncertain, how risk was controlled and who explicitly accepted the next action.

ABC DELIVERY

Give accurate, brief and clear information in logical order without removing essential limits or uncertainty..

- **Establish exact location, immediate danger, source and caller needs.**
- **Choose the communication mode, authorised response and explicit owner.**
- **Read back critical detail, record action and confirm safe handover.**

CONTROL PRINCIPLE

Use role-specific scripts as prompts, not substitutes for listening. Any emergency, safeguarding or safety-critical action must follow current controlled procedure.

ABC DELIVERY - COMMUNICATION PRACTICE

Use this page for simulation, coaching or review. Do not enter real personal data in an uncontrolled workbook.

- **Caller, channel and exact location:** ____________________________
- **What is happening and immediate danger:** _______________________
- **Words that acknowledged and clarified:** _________________________
- **Operational action, authority and owner:** _______________________
- **Read-back, update and handover:** ______________________________
- **Welfare and learning:** _______________________________________

PRACTICE

A strong call record shows what was known, what remained uncertain, how risk was controlled and who explicitly accepted the next action.

PHONETIC ALPHABET AND NUMBERS

Use authorised pronunciation and separate critical digits where procedures require..

- **Establish exact location, immediate danger, source and caller needs.**
- **Choose the communication mode, authorised response and explicit owner.**
- **Read back critical detail, record action and confirm safe handover.**

CONTROL PRINCIPLE

Use role-specific scripts as prompts, not substitutes for listening. Any emergency, safeguarding or safety-critical action must follow current controlled procedure.

PHONETIC ALPHABET AND NUMBERS - COMMUNICATION PRACTICE

Use this page for simulation, coaching or review. Do not enter real personal data in an uncontrolled workbook.

- **Caller, channel and exact location:** ________________________________
- **What is happening and immediate danger:** __________________________
- **Words that acknowledged and clarified:** ____________________________
- **Operational action, authority and owner:** __________________________
- **Read-back, update and handover:** _________________________________
- **Welfare and learning:** ____

PRACTICE

A strong call record shows what was known, what remained uncertain, how risk was controlled and who explicitly accepted the next action.

READ-BACK AND HEAR-BACK

The receiver repeats critical content and the sender checks and corrects it immediately..

- **Establish exact location, immediate danger, source and caller needs.**
- **Choose the communication mode, authorised response and explicit owner.**
- **Read back critical detail, record action and confirm safe handover.**

CONTROL PRINCIPLE

Use role-specific scripts as prompts, not substitutes for listening. Any emergency, safeguarding or safety-critical action must follow current controlled procedure.

READ-BACK AND HEAR-BACK - COMMUNICATION PRACTICE

Use this page for simulation, coaching or review. Do not enter real personal data in an uncontrolled workbook.

- **Caller, channel and exact location:** ______________________________
- **What is happening and immediate danger:** ________________________
- **Words that acknowledged and clarified:** ___________________________
- **Operational action, authority and owner:** _________________________
- **Read-back, update and handover:** ________________________________
- **Welfare and learning:** ____

PRACTICE

A strong call record shows what was known, what remained uncertain, how risk was controlled and who explicitly accepted the next action.

EMERGENCY PRIORITY

Use the emergency call route and priority language specified by current railway procedures..

- **Establish exact location, immediate danger, source and caller needs.**
- **Choose the communication mode, authorised response and explicit owner.**
- **Read back critical detail, record action and confirm safe handover.**

CONTROL PRINCIPLE

Use role-specific scripts as prompts, not substitutes for listening. Any emergency, safeguarding or safety-critical action must follow current controlled procedure.

EMERGENCY PRIORITY - COMMUNICATION PRACTICE

Use this page for simulation, coaching or review. Do not enter real personal data in an uncontrolled workbook.

- **Caller, channel and exact location:** ______________________________
- **What is happening and immediate danger:** ________________________
- **Words that acknowledged and clarified:** __________________________
- **Operational action, authority and owner:** ________________________
- **Read-back, update and handover:** ________________________________
- **Welfare and learning:** ____

PRACTICE

A strong call record shows what was known, what remained uncertain, how risk was controlled and who explicitly accepted the next action.

COMMUNICATION CHALLENGE

Stop and resolve ambiguity, incorrect read-back, conflicting authority or lost contact before proceeding..

- **Establish exact location, immediate danger, source and caller needs.**
- **Choose the communication mode, authorised response and explicit owner.**
- **Read back critical detail, record action and confirm safe handover.**

CONTROL PRINCIPLE

Use role-specific scripts as prompts, not substitutes for listening. Any emergency, safeguarding or safety-critical action must follow current controlled procedure.

COMMUNICATION CHALLENGE - COMMUNICATION PRACTICE

Use this page for simulation, coaching or review. Do not enter real personal data in an uncontrolled workbook.

- **Caller, channel and exact location:** ______________________________
- **What is happening and immediate danger:** ______________________________
- **Words that acknowledged and clarified:** ______________________________
- **Operational action, authority and owner:** ______________________________
- **Read-back, update and handover:** ______________________________
- **Welfare and learning:** ______________________________

PRACTICE

A strong call record shows what was known, what remained uncertain, how risk was controlled and who explicitly accepted the next action.

PART 9 | SWITCHING COMMUNICATION MODE

MATCH THE LANGUAGE TO THE RISK

A skilled controller can move between human support and formal operational protocol without losing either the person or the safe state.

IMMEDIATE MOVEMENT RISK

Use direct, short, closed-loop instruction through the authorised operational channel..

- **Establish exact location, immediate danger, source and caller needs.**
- **Choose the communication mode, authorised response and explicit owner.**
- **Read back critical detail, record action and confirm safe handover.**

CONTROL PRINCIPLE

Use role-specific scripts as prompts, not substitutes for listening. Any emergency, safeguarding or safety-critical action must follow current controlled procedure.

IMMEDIATE MOVEMENT RISK - COMMUNICATION PRACTICE

Use this page for simulation, coaching or review. Do not enter real personal data in an uncontrolled workbook.

- **Caller, channel and exact location:** ______________________________
- **What is happening and immediate danger:** ______________________________
- **Words that acknowledged and clarified:** ______________________________
- **Operational action, authority and owner:** ______________________________
- **Read-back, update and handover:** ______________________________
- **Welfare and learning:** ______________________________

> PRACTICE
>
> **A strong call record shows what was known, what remained uncertain, how risk was controlled and who explicitly accepted the next action.**

DISTRESSED PASSENGER

Use validating, paced questions while gathering location and danger information..

- **Establish exact location, immediate danger, source and caller needs.**
- **Choose the communication mode, authorised response and explicit owner.**
- **Read back critical detail, record action and confirm safe handover.**

CONTROL PRINCIPLE

Use role-specific scripts as prompts, not substitutes for listening. Any emergency, safeguarding or safety-critical action must follow current controlled procedure.

DISTRESSED PASSENGER - COMMUNICATION PRACTICE

Use this page for simulation, coaching or review. Do not enter real personal data in an uncontrolled workbook.

- **Caller, channel and exact location:** ______________________________
- **What is happening and immediate danger:** ______________________________
- **Words that acknowledged and clarified:** ______________________________
- **Operational action, authority and owner:** ______________________________
- **Read-back, update and handover:** ______________________________
- **Welfare and learning:** ______________________________

> **PRACTICE**
>
> **A strong call record shows what was known, what remained uncertain, how risk was controlled and who explicitly accepted the next action.**

AGGRESSIVE CALLER

Set behavioural boundaries, identify the actionable problem and preserve evidence of credible threat..

- **Establish exact location, immediate danger, source and caller needs.**
- **Choose the communication mode, authorised response and explicit owner.**
- **Read back critical detail, record action and confirm safe handover.**

CONTROL PRINCIPLE

Use role-specific scripts as prompts, not substitutes for listening. Any emergency, safeguarding or safety-critical action must follow current controlled procedure.

AGGRESSIVE CALLER - COMMUNICATION PRACTICE

Use this page for simulation, coaching or review. Do not enter real personal data in an uncontrolled workbook.

- **Caller, channel and exact location:** ______________________________
- **What is happening and immediate danger:** ______________________________
- **Words that acknowledged and clarified:** ______________________________
- **Operational action, authority and owner:** ______________________________
- **Read-back, update and handover:** ______________________________
- **Welfare and learning:** ______________________________

PRACTICE

A strong call record shows what was known, what remained uncertain, how risk was controlled and who explicitly accepted the next action.

STAFF UNCERTAINTY

Reinforce stopping and checking, establish the safest state and connect the correct authority..

- **Establish exact location, immediate danger, source and caller needs.**
- **Choose the communication mode, authorised response and explicit owner.**
- **Read back critical detail, record action and confirm safe handover.**

CONTROL PRINCIPLE

Use role-specific scripts as prompts, not substitutes for listening. Any emergency, safeguarding or safety-critical action must follow current controlled procedure.

STAFF UNCERTAINTY - COMMUNICATION PRACTICE

Use this page for simulation, coaching or review. Do not enter real personal data in an uncontrolled workbook.

- **Caller, channel and exact location:** ______________________________
- **What is happening and immediate danger:** ________________________
- **Words that acknowledged and clarified:** __________________________
- **Operational action, authority and owner:** ________________________
- **Read-back, update and handover:** ________________________________
- **Welfare and learning:** ____

PRACTICE

A strong call record shows what was known, what remained uncertain, how risk was controlled and who explicitly accepted the next action.

MULTI-AGENCY RESPONSE

Use structured situation, impact, action, command and request updates..

- **Establish exact location, immediate danger, source and caller needs.**
- **Choose the communication mode, authorised response and explicit owner.**
- **Read back critical detail, record action and confirm safe handover.**

CONTROL PRINCIPLE

Use role-specific scripts as prompts, not substitutes for listening. Any emergency, safeguarding or safety-critical action must follow current controlled procedure.

MULTI-AGENCY RESPONSE - COMMUNICATION PRACTICE

Use this page for simulation, coaching or review. Do not enter real personal data in an uncontrolled workbook.

- **Caller, channel and exact location:** ______________________________
- **What is happening and immediate danger:** ______________________________
- **Words that acknowledged and clarified:** ______________________________
- **Operational action, authority and owner:** ______________________________
- **Read-back, update and handover:** ______________________________
- **Welfare and learning:** ______________________________

PRACTICE

A strong call record shows what was known, what remained uncertain, how risk was controlled and who explicitly accepted the next action.

RETURNING TO NORMAL

Explain the change in pace and confirm the person still understands the next step and ownership..

- **Establish exact location, immediate danger, source and caller needs.**
- **Choose the communication mode, authorised response and explicit owner.**
- **Read back critical detail, record action and confirm safe handover.**

CONTROL PRINCIPLE

Use role-specific scripts as prompts, not substitutes for listening. Any emergency, safeguarding or safety-critical action must follow current controlled procedure.

RETURNING TO NORMAL - COMMUNICATION PRACTICE

Use this page for simulation, coaching or review. Do not enter real personal data in an uncontrolled workbook.

- **Caller, channel and exact location:** ______________________________
- **What is happening and immediate danger:** ________________________
- **Words that acknowledged and clarified:** __________________________
- **Operational action, authority and owner:** ________________________
- **Read-back, update and handover:** _______________________________
- **Welfare and learning:** ____

PRACTICE

A strong call record shows what was known, what remained uncertain, how risk was controlled and who explicitly accepted the next action.

PART 10 | PASSENGER ASSIST

RELIABLE ASSISTANCE IS A CONTROL CHAIN

Passenger Assist depends on accurate records, accepted handovers and recovery when any link fails.

THE ASSISTANCE REQUEST

Record journey, assistance required, mobility equipment, contact, communication need and relevant consent..

- **Establish exact location, immediate danger, source and caller needs.**
- **Choose the communication mode, authorised response and explicit owner.**
- **Read back critical detail, record action and confirm safe handover.**

CONTROL PRINCIPLE

Use role-specific scripts as prompts, not substitutes for listening. Any emergency, safeguarding or safety-critical action must follow current controlled procedure.

THE ASSISTANCE REQUEST - COMMUNICATION PRACTICE

Use this page for simulation, coaching or review. Do not enter real personal data in an uncontrolled workbook.

- **Caller, channel and exact location:** ________________________________
- **What is happening and immediate danger:** __________________________
- **Words that acknowledged and clarified:** ____________________________
- **Operational action, authority and owner:** ___________________________
- **Read-back, update and handover:** __________________________________
- **Welfare and learning:** ____

PRACTICE

A strong call record shows what was known, what remained uncertain, how risk was controlled and who explicitly accepted the next action.

DEPARTURE ACCEPTANCE

Confirm responsible staff, meeting location, train and contingency rather than assuming a system entry creates delivery..

- **Establish exact location, immediate danger, source and caller needs.**
- **Choose the communication mode, authorised response and explicit owner.**
- **Read back critical detail, record action and confirm safe handover.**

CONTROL PRINCIPLE

Use role-specific scripts as prompts, not substitutes for listening. Any emergency, safeguarding or safety-critical action must follow current controlled procedure.

DEPARTURE ACCEPTANCE - COMMUNICATION PRACTICE

Use this page for simulation, coaching or review. Do not enter real personal data in an uncontrolled workbook.

- **Caller, channel and exact location:** ______________________________
- **What is happening and immediate danger:** ______________________________
- **Words that acknowledged and clarified:** ______________________________
- **Operational action, authority and owner:** ______________________________
- **Read-back, update and handover:** ______________________________
- **Welfare and learning:** ______________________________

PRACTICE

A strong call record shows what was known, what remained uncertain, how risk was controlled and who explicitly accepted the next action.

TRAIN AND COACH DETAIL

Use service, formation, coach, door and accessible-space information available from current systems..

- **Establish exact location, immediate danger, source and caller needs.**
- **Choose the communication mode, authorised response and explicit owner.**
- **Read back critical detail, record action and confirm safe handover.**

CONTROL PRINCIPLE

Use role-specific scripts as prompts, not substitutes for listening. Any emergency, safeguarding or safety-critical action must follow current controlled procedure.

TRAIN AND COACH DETAIL - COMMUNICATION PRACTICE

Use this page for simulation, coaching or review. Do not enter real personal data in an uncontrolled workbook.

- **Caller, channel and exact location:** ________________________________
- **What is happening and immediate danger:** __________________________
- **Words that acknowledged and clarified:** ____________________________
- **Operational action, authority and owner:** __________________________
- **Read-back, update and handover:** __________________________________
- **Welfare and learning:** ____

PRACTICE

A strong call record shows what was known, what remained uncertain, how risk was controlled and who explicitly accepted the next action.

DESTINATION HANDOVER

Transmit expected arrival, location and required action and obtain positive acceptance..

- **Establish exact location, immediate danger, source and caller needs.**
- **Choose the communication mode, authorised response and explicit owner.**
- **Read back critical detail, record action and confirm safe handover.**

CONTROL PRINCIPLE

Use role-specific scripts as prompts, not substitutes for listening. Any emergency, safeguarding or safety-critical action must follow current controlled procedure.

DESTINATION HANDOVER - COMMUNICATION PRACTICE

Use this page for simulation, coaching or review. Do not enter real personal data in an uncontrolled workbook.

- **Caller, channel and exact location:** ________________________________
- **What is happening and immediate danger:** __________________________
- **Words that acknowledged and clarified:** ____________________________
- **Operational action, authority and owner:** __________________________
- **Read-back, update and handover:** ___________________________________
- **Welfare and learning:** ____

PRACTICE

A strong call record shows what was known, what remained uncertain, how risk was controlled and who explicitly accepted the next action.

TURN-UP-AND-GO

Support unbooked assistance through the operator's Accessible Travel Policy and operational capability..

- **Establish exact location, immediate danger, source and caller needs.**
- **Choose the communication mode, authorised response and explicit owner.**
- **Read back critical detail, record action and confirm safe handover.**

CONTROL PRINCIPLE

Use role-specific scripts as prompts, not substitutes for listening. Any emergency, safeguarding or safety-critical action must follow current controlled procedure.

TURN-UP-AND-GO - COMMUNICATION PRACTICE

Use this page for simulation, coaching or review. Do not enter real personal data in an uncontrolled workbook.

- **Caller, channel and exact location:** ________________________________
- **What is happening and immediate danger:** ____________________________
- **Words that acknowledged and clarified:** _____________________________
- **Operational action, authority and owner:** ____________________________
- **Read-back, update and handover:** ___________________________________
- **Welfare and learning:** ____

PRACTICE

A strong call record shows what was known, what remained uncertain, how risk was controlled and who explicitly accepted the next action.

COMPLETION AND LEARNING

Confirm assistance happened, record failure or variation and feed evidence into continuous improvement..

- **Establish exact location, immediate danger, source and caller needs.**
- **Choose the communication mode, authorised response and explicit owner.**
- **Read back critical detail, record action and confirm safe handover.**

CONTROL PRINCIPLE

Use role-specific scripts as prompts, not substitutes for listening. Any emergency, safeguarding or safety-critical action must follow current controlled procedure.

COMPLETION AND LEARNING - COMMUNICATION PRACTICE

Use this page for simulation, coaching or review. Do not enter real personal data in an uncontrolled workbook.

- **Caller, channel and exact location:** ______________________________
- **What is happening and immediate danger:** ________________________
- **Words that acknowledged and clarified:** __________________________
- **Operational action, authority and owner:** ________________________
- **Read-back, update and handover:** ________________________________
- **Welfare and learning:** ____

PRACTICE

A strong call record shows what was known, what remained uncertain, how risk was controlled and who explicitly accepted the next action.

PART 11 | ASSISTANCE FAILURE

RECOVER THE PERSON, THEN INVESTIGATE THE SYSTEM

A booking record does not prove that a passenger is safe, supported or at the location the system predicts.

SAFETY BEFORE BOOKING HISTORY

Establish where the passenger is, whether they can disembark and whether medical, personal-care or onward risks exist..

- **Establish exact location, immediate danger, source and caller needs.**
- **Choose the communication mode, authorised response and explicit owner.**
- **Read back critical detail, record action and confirm safe handover.**

CONTROL PRINCIPLE

Use role-specific scripts as prompts, not substitutes for listening. Any emergency, safeguarding or safety-critical action must follow current controlled procedure.

SAFETY BEFORE BOOKING HISTORY - COMMUNICATION PRACTICE

Use this page for simulation, coaching or review. Do not enter real personal data in an uncontrolled workbook.

- **Caller, channel and exact location:** ________________________________
- **What is happening and immediate danger:** ________________________________
- **Words that acknowledged and clarified:** ________________________________
- **Operational action, authority and owner:** ________________________________
- **Read-back, update and handover:** ________________________________
- **Welfare and learning:** ________________________________

PRACTICE

A strong call record shows what was known, what remained uncertain, how risk was controlled and who explicitly accepted the next action.

ACKNOWLEDGE FAILURE

Apologise for the missed assistance without arguing about which team or system caused it..

- **Establish exact location, immediate danger, source and caller needs.**
- **Choose the communication mode, authorised response and explicit owner.**
- **Read back critical detail, record action and confirm safe handover.**

CONTROL PRINCIPLE

Use role-specific scripts as prompts, not substitutes for listening. Any emergency, safeguarding or safety-critical action must follow current controlled procedure.

ACKNOWLEDGE FAILURE - COMMUNICATION PRACTICE

Use this page for simulation, coaching or review. Do not enter real personal data in an uncontrolled workbook.

- **Caller, channel and exact location:** ______________________________
- **What is happening and immediate danger:** ______________________________
- **Words that acknowledged and clarified:** ______________________________
- **Operational action, authority and owner:** ______________________________
- **Read-back, update and handover:** ______________________________
- **Welfare and learning:** ______________________________

PRACTICE

A strong call record shows what was known, what remained uncertain, how risk was controlled and who explicitly accepted the next action.

IMMEDIATE RECOVERY

Find competent local support, accessible route and realistic next journey while maintaining contact..

- **Establish exact location, immediate danger, source and caller needs.**
- **Choose the communication mode, authorised response and explicit owner.**
- **Read back critical detail, record action and confirm safe handover.**

CONTROL PRINCIPLE

Use role-specific scripts as prompts, not substitutes for listening. Any emergency, safeguarding or safety-critical action must follow current controlled procedure.

IMMEDIATE RECOVERY - COMMUNICATION PRACTICE

Use this page for simulation, coaching or review. Do not enter real personal data in an uncontrolled workbook.

- **Caller, channel and exact location: ______________________________**
- **What is happening and immediate danger: ________________________**
- **Words that acknowledged and clarified: __________________________**
- **Operational action, authority and owner: ________________________**
- **Read-back, update and handover: _______________________________**
- **Welfare and learning: ____**

PRACTICE

A strong call record shows what was known, what remained uncertain, how risk was controlled and who explicitly accepted the next action.

OWNERSHIP AND UPDATE

Give a named or functional owner and a definite update time even when resolution is still pending..

- **Establish exact location, immediate danger, source and caller needs.**
- **Choose the communication mode, authorised response and explicit owner.**
- **Read back critical detail, record action and confirm safe handover.**

CONTROL PRINCIPLE

Use role-specific scripts as prompts, not substitutes for listening. Any emergency, safeguarding or safety-critical action must follow current controlled procedure.

OWNERSHIP AND UPDATE - COMMUNICATION PRACTICE

Use this page for simulation, coaching or review. Do not enter real personal data in an uncontrolled workbook.

- **Caller, channel and exact location: ________________________________**
- **What is happening and immediate danger: ________________________________**
- **Words that acknowledged and clarified: ________________________________**
- **Operational action, authority and owner: ________________________________**
- **Read-back, update and handover: ________________________________**
- **Welfare and learning: ________________________________**

PRACTICE

A strong call record shows what was known, what remained uncertain, how risk was controlled and who explicitly accepted the next action.

FAILED STATION HANDOVER

Escalate non-acceptance, changed train or unstaffed destination before the passenger becomes stranded..

- **Establish exact location, immediate danger, source and caller needs.**
- **Choose the communication mode, authorised response and explicit owner.**
- **Read back critical detail, record action and confirm safe handover.**

CONTROL PRINCIPLE

Use role-specific scripts as prompts, not substitutes for listening. Any emergency, safeguarding or safety-critical action must follow current controlled procedure.

FAILED STATION HANDOVER - COMMUNICATION PRACTICE

Use this page for simulation, coaching or review. Do not enter real personal data in an uncontrolled workbook.

- **Caller, channel and exact location: ______________________________**
- **What is happening and immediate danger: ________________________**
- **Words that acknowledged and clarified: __________________________**
- **Operational action, authority and owner: ________________________**
- **Read-back, update and handover: _______________________________**
- **Welfare and learning: ______________________________________**

PRACTICE

A strong call record shows what was known, what remained uncertain, how risk was controlled and who explicitly accepted the next action.

POST-EVENT REVIEW

Examine booking, system, communication, staffing, train change and acceptance evidence rather than blaming the final person in the chain..

- **Establish exact location, immediate danger, source and caller needs.**
- **Choose the communication mode, authorised response and explicit owner.**
- **Read back critical detail, record action and confirm safe handover.**

CONTROL PRINCIPLE

Use role-specific scripts as prompts, not substitutes for listening. Any emergency, safeguarding or safety-critical action must follow current controlled procedure.

POST-EVENT REVIEW - COMMUNICATION PRACTICE

Use this page for simulation, coaching or review. Do not enter real personal data in an uncontrolled workbook.

- **Caller, channel and exact location:** ______________________________
- **What is happening and immediate danger:** __________________________
- **Words that acknowledged and clarified:** ___________________________
- **Operational action, authority and owner:** __________________________
- **Read-back, update and handover:** _________________________________
- **Welfare and learning:** ____

PRACTICE

A strong call record shows what was known, what remained uncertain, how risk was controlled and who explicitly accepted the next action.

PART 12 | ACCESSIBLE COMMUNICATION

CHANGE THE CHANNEL, NOT THE DIGNITY

People communicate differently; the controller must remove barriers without assuming inability or speaking only to a companion.

HEARING AND SPEECH DIFFERENCES

Use text, relay, slower questions or confirmation agreed with the passenger and allow time..

- **Establish exact location, immediate danger, source and caller needs.**
- **Choose the communication mode, authorised response and explicit owner.**
- **Read back critical detail, record action and confirm safe handover.**

CONTROL PRINCIPLE

Use role-specific scripts as prompts, not substitutes for listening. Any emergency, safeguarding or safety-critical action must follow current controlled procedure.

HEARING AND SPEECH DIFFERENCES - COMMUNICATION PRACTICE

Use this page for simulation, coaching or review. Do not enter real personal data in an uncontrolled workbook.

- **Caller, channel and exact location:** ______________________________
- **What is happening and immediate danger:** ________________________
- **Words that acknowledged and clarified:** __________________________
- **Operational action, authority and owner:** ________________________
- **Read-back, update and handover:** _______________________________
- **Welfare and learning:** ____

PRACTICE

A strong call record shows what was known, what remained uncertain, how risk was controlled and who explicitly accepted the next action.

VISUAL IMPAIRMENT

Describe location and action precisely; do not rely on colour, gesture or screen-only information..

- **Establish exact location, immediate danger, source and caller needs.**
- **Choose the communication mode, authorised response and explicit owner.**
- **Read back critical detail, record action and confirm safe handover.**

CONTROL PRINCIPLE

Use role-specific scripts as prompts, not substitutes for listening. Any emergency, safeguarding or safety-critical action must follow current controlled procedure.

VISUAL IMPAIRMENT - COMMUNICATION PRACTICE

Use this page for simulation, coaching or review. Do not enter real personal data in an uncontrolled workbook.

- **Caller, channel and exact location:** ______________________________
- **What is happening and immediate danger:** _________________________
- **Words that acknowledged and clarified:** ___________________________
- **Operational action, authority and owner:** _________________________
- **Read-back, update and handover:** ________________________________
- **Welfare and learning:** ____

> **PRACTICE**
>
> **A strong call record shows what was known, what remained uncertain, how risk was controlled and who explicitly accepted the next action.**

LEARNING DISABILITY

Use concrete language, one step at a time and a genuine check of understanding..

- **Establish exact location, immediate danger, source and caller needs.**
- **Choose the communication mode, authorised response and explicit owner.**
- **Read back critical detail, record action and confirm safe handover.**

CONTROL PRINCIPLE

Use role-specific scripts as prompts, not substitutes for listening. Any emergency, safeguarding or safety-critical action must follow current controlled procedure.

LEARNING DISABILITY - COMMUNICATION PRACTICE

Use this page for simulation, coaching or review. Do not enter real personal data in an uncontrolled workbook.

- **Caller, channel and exact location:** ______________________________
- **What is happening and immediate danger:** ______________________________
- **Words that acknowledged and clarified:** ______________________________
- **Operational action, authority and owner:** ______________________________
- **Read-back, update and handover:** ______________________________
- **Welfare and learning:** ______________________________

PRACTICE

A strong call record shows what was known, what remained uncertain, how risk was controlled and who explicitly accepted the next action.

NEURODIVERGENCE AND DISTRESS

Reduce unnecessary language and sensory demand while respecting the person's stated communication preference..

- **Establish exact location, immediate danger, source and caller needs.**
- **Choose the communication mode, authorised response and explicit owner.**
- **Read back critical detail, record action and confirm safe handover.**

CONTROL PRINCIPLE

Use role-specific scripts as prompts, not substitutes for listening. Any emergency, safeguarding or safety-critical action must follow current controlled procedure.

NEURODIVERGENCE AND DISTRESS - COMMUNICATION PRACTICE

Use this page for simulation, coaching or review. Do not enter real personal data in an uncontrolled workbook.

- **Caller, channel and exact location: ______________________________**
- **What is happening and immediate danger: ________________________**
- **Words that acknowledged and clarified: __________________________**
- **Operational action, authority and owner: ________________________**
- **Read-back, update and handover: _______________________________**
- **Welfare and learning: ____**

> **PRACTICE**
>
> **A strong call record shows what was known, what remained uncertain, how risk was controlled and who explicitly accepted the next action.**

LIMITED ENGLISH

Use authorised interpretation or accessible alternative and verify essential safety information..

- **Establish exact location, immediate danger, source and caller needs.**
- **Choose the communication mode, authorised response and explicit owner.**
- **Read back critical detail, record action and confirm safe handover.**

CONTROL PRINCIPLE

Use role-specific scripts as prompts, not substitutes for listening. Any emergency, safeguarding or safety-critical action must follow current controlled procedure.

LIMITED ENGLISH - COMMUNICATION PRACTICE

Use this page for simulation, coaching or review. Do not enter real personal data in an uncontrolled workbook.

- **Caller, channel and exact location:** ______________________________
- **What is happening and immediate danger:** ______________________________
- **Words that acknowledged and clarified:** ______________________________
- **Operational action, authority and owner:** ______________________________
- **Read-back, update and handover:** ______________________________
- **Welfare and learning:** ______________________________

PRACTICE

A strong call record shows what was known, what remained uncertain, how risk was controlled and who explicitly accepted the next action.

COMPANIONS AND CONSENT

Speak to the passenger, confirm what may be shared and involve companions without erasing the passenger's choice..

- **Establish exact location, immediate danger, source and caller needs.**
- **Choose the communication mode, authorised response and explicit owner.**
- **Read back critical detail, record action and confirm safe handover.**

CONTROL PRINCIPLE

Use role-specific scripts as prompts, not substitutes for listening. Any emergency, safeguarding or safety-critical action must follow current controlled procedure.

COMPANIONS AND CONSENT - COMMUNICATION PRACTICE

Use this page for simulation, coaching or review. Do not enter real personal data in an uncontrolled workbook.

- **Caller, channel and exact location:** ______________________________
- **What is happening and immediate danger:** ________________________
- **Words that acknowledged and clarified:** __________________________
- **Operational action, authority and owner:** ________________________
- **Read-back, update and handover:** ________________________________
- **Welfare and learning:** ____

PRACTICE

A strong call record shows what was known, what remained uncertain, how risk was controlled and who explicitly accepted the next action.

VULNERABILITY REQUIRES CONTROLLED CURIOSITY

Controllers may be the first to notice that a railway problem is also a safeguarding concern.

MISSING OR SEPARATED CHILD

Establish exact identity, location, safe supervision and police or safeguarding escalation without broadcasting unnecessary detail..

- **Establish exact location, immediate danger, source and caller needs.**
- **Choose the communication mode, authorised response and explicit owner.**
- **Read back critical detail, record action and confirm safe handover.**

CONTROL PRINCIPLE

Use role-specific scripts as prompts, not substitutes for listening. Any emergency, safeguarding or safety-critical action must follow current controlled procedure.

MISSING OR SEPARATED CHILD - COMMUNICATION PRACTICE

Use this page for simulation, coaching or review. Do not enter real personal data in an uncontrolled workbook.

- **Caller, channel and exact location:** ______________________________
- **What is happening and immediate danger:** ________________________
- **Words that acknowledged and clarified:** ___________________________
- **Operational action, authority and owner:** _________________________
- **Read-back, update and handover:** _______________________________
- **Welfare and learning:** ____

PRACTICE

A strong call record shows what was known, what remained uncertain, how risk was controlled and who explicitly accepted the next action.

CONFUSED ADULT

Consider health, dementia, learning need, intoxication and exploitation while arranging safe local contact..

- **Establish exact location, immediate danger, source and caller needs.**
- **Choose the communication mode, authorised response and explicit owner.**
- **Read back critical detail, record action and confirm safe handover.**

CONTROL PRINCIPLE

Use role-specific scripts as prompts, not substitutes for listening. Any emergency, safeguarding or safety-critical action must follow current controlled procedure.

CONFUSED ADULT - COMMUNICATION PRACTICE

Use this page for simulation, coaching or review. Do not enter real personal data in an uncontrolled workbook.

- **Caller, channel and exact location:** ______________________________
- **What is happening and immediate danger:** ______________________________
- **Words that acknowledged and clarified:** ______________________________
- **Operational action, authority and owner:** ______________________________
- **Read-back, update and handover:** ______________________________
- **Welfare and learning:** ______________________________

PRACTICE

A strong call record shows what was known, what remained uncertain, how risk was controlled and who explicitly accepted the next action.

DOMESTIC ABUSE AND EXPLOITATION

Use discreet questions where safe, do not alert a suspected abuser and follow the specialist escalation route..

- **Establish exact location, immediate danger, source and caller needs.**
- **Choose the communication mode, authorised response and explicit owner.**
- **Read back critical detail, record action and confirm safe handover.**

CONTROL PRINCIPLE

Use role-specific scripts as prompts, not substitutes for listening. Any emergency, safeguarding or safety-critical action must follow current controlled procedure.

DOMESTIC ABUSE AND EXPLOITATION - COMMUNICATION PRACTICE

Use this page for simulation, coaching or review. Do not enter real personal data in an uncontrolled workbook.

- **Caller, channel and exact location:** ______________________________
- **What is happening and immediate danger:** ________________________
- **Words that acknowledged and clarified:** __________________________
- **Operational action, authority and owner:** ________________________
- **Read-back, update and handover:** ________________________________
- **Welfare and learning:** ____

PRACTICE

A strong call record shows what was known, what remained uncertain, how risk was controlled and who explicitly accepted the next action.

STRANDED PASSENGER

Assess time, location, accessibility, funds, medication, vulnerability and safe onward options..

- **Establish exact location, immediate danger, source and caller needs.**
- **Choose the communication mode, authorised response and explicit owner.**
- **Read back critical detail, record action and confirm safe handover.**

CONTROL PRINCIPLE

Use role-specific scripts as prompts, not substitutes for listening. Any emergency, safeguarding or safety-critical action must follow current controlled procedure.

STRANDED PASSENGER - COMMUNICATION PRACTICE

Use this page for simulation, coaching or review. Do not enter real personal data in an uncontrolled workbook.

- **Caller, channel and exact location:** ______________________________
- **What is happening and immediate danger:** ______________________________
- **Words that acknowledged and clarified:** ______________________________
- **Operational action, authority and owner:** ______________________________
- **Read-back, update and handover:** ______________________________
- **Welfare and learning:** ______________________________

PRACTICE

A strong call record shows what was known, what remained uncertain, how risk was controlled and who explicitly accepted the next action.

HATE CRIME AND HARASSMENT

Take the report seriously, protect immediate safety, preserve choice and arrange BTP or emergency support..

- **Establish exact location, immediate danger, source and caller needs.**
- **Choose the communication mode, authorised response and explicit owner.**
- **Read back critical detail, record action and confirm safe handover.**

CONTROL PRINCIPLE

Use role-specific scripts as prompts, not substitutes for listening. Any emergency, safeguarding or safety-critical action must follow current controlled procedure.

HATE CRIME AND HARASSMENT - COMMUNICATION PRACTICE

Use this page for simulation, coaching or review. Do not enter real personal data in an uncontrolled workbook.

- **Caller, channel and exact location: ______________________________**
- **What is happening and immediate danger: ________________________**
- **Words that acknowledged and clarified: ___________________________**
- **Operational action, authority and owner: _________________________**
- **Read-back, update and handover: _________________________________**
- **Welfare and learning: ____**

PRACTICE

A strong call record shows what was known, what remained uncertain, how risk was controlled and who explicitly accepted the next action.

CONFIDENTIALITY LIMITS

Share necessary information with authorised responders and never promise secrecy where safety or safeguarding requires action..

- **Establish exact location, immediate danger, source and caller needs.**
- **Choose the communication mode, authorised response and explicit owner.**
- **Read back critical detail, record action and confirm safe handover.**

CONTROL PRINCIPLE

Use role-specific scripts as prompts, not substitutes for listening. Any emergency, safeguarding or safety-critical action must follow current controlled procedure.

CONFIDENTIALITY LIMITS - COMMUNICATION PRACTICE

Use this page for simulation, coaching or review. Do not enter real personal data in an uncontrolled workbook.

- **Caller, channel and exact location:** ______________________________
- **What is happening and immediate danger:** ________________________
- **Words that acknowledged and clarified:** __________________________
- **Operational action, authority and owner:** ________________________
- **Read-back, update and handover:** _________________________________
- **Welfare and learning:** ____

PRACTICE

A strong call record shows what was known, what remained uncertain, how risk was controlled and who explicitly accepted the next action.

VALIDATION DOES NOT REQUIRE ACCEPTING ABUSE

Controllers can acknowledge anger, set boundaries and preserve personal safety while still addressing the actionable issue.

FRUSTRATION OR THREAT

Distinguish emotional language, targeted abuse, credible threat, violence in progress and emergency danger..

- **Establish exact location, immediate danger, source and caller needs.**
- **Choose the communication mode, authorised response and explicit owner.**
- **Read back critical detail, record action and confirm safe handover.**

CONTROL PRINCIPLE

Use role-specific scripts as prompts, not substitutes for listening. Any emergency, safeguarding or safety-critical action must follow current controlled procedure.

FRUSTRATION OR THREAT - COMMUNICATION PRACTICE

Use this page for simulation, coaching or review. Do not enter real personal data in an uncontrolled workbook.

- **Caller, channel and exact location: ________________________________**
- **What is happening and immediate danger: ____________________________**
- **Words that acknowledged and clarified: ______________________________**
- **Operational action, authority and owner: ____________________________**
- **Read-back, update and handover: ___________________________________**
- **Welfare and learning: ____**

PRACTICE

A strong call record shows what was known, what remained uncertain, how risk was controlled and who explicitly accepted the next action.

SET A CLEAR BOUNDARY

State that help can continue but threats, discriminatory abuse or repeated personal attack cannot..

- **Establish exact location, immediate danger, source and caller needs.**
- **Choose the communication mode, authorised response and explicit owner.**
- **Read back critical detail, record action and confirm safe handover.**

CONTROL PRINCIPLE

Use role-specific scripts as prompts, not substitutes for listening. Any emergency, safeguarding or safety-critical action must follow current controlled procedure.

SET A CLEAR BOUNDARY - COMMUNICATION PRACTICE

Use this page for simulation, coaching or review. Do not enter real personal data in an uncontrolled workbook.

- **Caller, channel and exact location:** ______________________________
- **What is happening and immediate danger:** ________________________
- **Words that acknowledged and clarified:** __________________________
- **Operational action, authority and owner:** ________________________
- **Read-back, update and handover:** _______________________________
- **Welfare and learning:** ____

PRACTICE

A strong call record shows what was known, what remained uncertain, how risk was controlled and who explicitly accepted the next action.

OFFER CONTROLLED CHOICES

Provide realistic options within authority rather than bargaining with unsafe or impossible demands..

- **Establish exact location, immediate danger, source and caller needs.**
- **Choose the communication mode, authorised response and explicit owner.**
- **Read back critical detail, record action and confirm safe handover.**

CONTROL PRINCIPLE

Use role-specific scripts as prompts, not substitutes for listening. Any emergency, safeguarding or safety-critical action must follow current controlled procedure.

OFFER CONTROLLED CHOICES - COMMUNICATION PRACTICE

Use this page for simulation, coaching or review. Do not enter real personal data in an uncontrolled workbook.

- **Caller, channel and exact location: ______________________**
- **What is happening and immediate danger: ____________________**
- **Words that acknowledged and clarified: _____________________**
- **Operational action, authority and owner: ___________________**
- **Read-back, update and handover: _________________________**
- **Welfare and learning: ________________________________**

PRACTICE

A strong call record shows what was known, what remained uncertain, how risk was controlled and who explicitly accepted the next action.

ENDING A CALL

Follow policy, warn where appropriate, preserve emergency information and arrange escalation before disconnecting..

- **Establish exact location, immediate danger, source and caller needs.**
- **Choose the communication mode, authorised response and explicit owner.**
- **Read back critical detail, record action and confirm safe handover.**

CONTROL PRINCIPLE

Use role-specific scripts as prompts, not substitutes for listening. Any emergency, safeguarding or safety-critical action must follow current controlled procedure.

ENDING A CALL - COMMUNICATION PRACTICE

Use this page for simulation, coaching or review. Do not enter real personal data in an uncontrolled workbook.

- **Caller, channel and exact location:** ______________________________
- **What is happening and immediate danger:** ________________________
- **Words that acknowledged and clarified:** __________________________
- **Operational action, authority and owner:** ________________________
- **Read-back, update and handover:** _______________________________
- **Welfare and learning:** ____

PRACTICE

A strong call record shows what was known, what remained uncertain, how risk was controlled and who explicitly accepted the next action.

THREAT EVIDENCE

Record exact words, caller details, location, target and immediacy without embellishment..

- **Establish exact location, immediate danger, source and caller needs.**
- **Choose the communication mode, authorised response and explicit owner.**
- **Read back critical detail, record action and confirm safe handover.**

CONTROL PRINCIPLE

Use role-specific scripts as prompts, not substitutes for listening. Any emergency, safeguarding or safety-critical action must follow current controlled procedure.

THREAT EVIDENCE - COMMUNICATION PRACTICE

Use this page for simulation, coaching or review. Do not enter real personal data in an uncontrolled workbook.

- **Caller, channel and exact location:** ________________________________
- **What is happening and immediate danger:** ____________________________
- **Words that acknowledged and clarified:** ____________________________
- **Operational action, authority and owner:** __________________________
- **Read-back, update and handover:** __________________________________
- **Welfare and learning:** ____

PRACTICE

A strong call record shows what was known, what remained uncertain, how risk was controlled and who explicitly accepted the next action.

CONTROLLER AFTERCARE

Allow recovery, supervisor support and police or welfare follow-up rather than sending the controller straight to another call..

- **Establish exact location, immediate danger, source and caller needs.**
- **Choose the communication mode, authorised response and explicit owner.**
- **Read back critical detail, record action and confirm safe handover.**

CONTROL PRINCIPLE

Use role-specific scripts as prompts, not substitutes for listening. Any emergency, safeguarding or safety-critical action must follow current controlled procedure.

CONTROLLER AFTERCARE - COMMUNICATION PRACTICE

Use this page for simulation, coaching or review. Do not enter real personal data in an uncontrolled workbook.

- **Caller, channel and exact location:** ________________________________
- **What is happening and immediate danger:** __________________________
- **Words that acknowledged and clarified:** ____________________________
- **Operational action, authority and owner:** __________________________
- **Read-back, update and handover:** _________________________________
- **Welfare and learning:** ____

PRACTICE

A strong call record shows what was known, what remained uncertain, how risk was controlled and who explicitly accepted the next action.

PART 15 | STAFF CALLING CONTROL

MAKE IT SAFE TO REPORT UNCERTAINTY

The controller's response determines whether staff disclose error, fatigue, fear and doubt early or hide them until risk grows.

STAFF UNDER PRESSURE

Ask for location, safe state, immediate hazard and help required before questioning performance..

- **Establish exact location, immediate danger, source and caller needs.**
- **Choose the communication mode, authorised response and explicit owner.**
- **Read back critical detail, record action and confirm safe handover.**

CONTROL PRINCIPLE

Use role-specific scripts as prompts, not substitutes for listening. Any emergency, safeguarding or safety-critical action must follow current controlled procedure.

STAFF UNDER PRESSURE - COMMUNICATION PRACTICE

Use this page for simulation, coaching or review. Do not enter real personal data in an uncontrolled workbook.

- **Caller, channel and exact location: ______________________________**
- **What is happening and immediate danger: ________________________**
- **Words that acknowledged and clarified: __________________________**
- **Operational action, authority and owner: ________________________**
- **Read-back, update and handover: _______________________________**
- **Welfare and learning: ______________________________________**

PRACTICE

A strong call record shows what was known, what remained uncertain, how risk was controlled and who explicitly accepted the next action.

REPORTING A MISTAKE

Protect people, contain consequence and gather facts; accountability and investigation follow after the safe state..

- **Establish exact location, immediate danger, source and caller needs.**
- **Choose the communication mode, authorised response and explicit owner.**
- **Read back critical detail, record action and confirm safe handover.**

CONTROL PRINCIPLE

Use role-specific scripts as prompts, not substitutes for listening. Any emergency, safeguarding or safety-critical action must follow current controlled procedure.

REPORTING A MISTAKE - COMMUNICATION PRACTICE

Use this page for simulation, coaching or review. Do not enter real personal data in an uncontrolled workbook.

- **Caller, channel and exact location:** ______________________________
- **What is happening and immediate danger:** ________________________
- **Words that acknowledged and clarified:** __________________________
- **Operational action, authority and owner:** ________________________
- **Read-back, update and handover:** _______________________________
- **Welfare and learning:** ____

PRACTICE

A strong call record shows what was known, what remained uncertain, how risk was controlled and who explicitly accepted the next action.

RULE UNCERTAINTY

Praise stopping and checking, connect the competent authority and do not improvise an operational rule..

- **Establish exact location, immediate danger, source and caller needs.**
- **Choose the communication mode, authorised response and explicit owner.**
- **Read back critical detail, record action and confirm safe handover.**

CONTROL PRINCIPLE

Use role-specific scripts as prompts, not substitutes for listening. Any emergency, safeguarding or safety-critical action must follow current controlled procedure.

RULE UNCERTAINTY - COMMUNICATION PRACTICE

Use this page for simulation, coaching or review. Do not enter real personal data in an uncontrolled workbook.

- **Caller, channel and exact location:** ______________________________
- **What is happening and immediate danger:** ________________________
- **Words that acknowledged and clarified:** __________________________
- **Operational action, authority and owner:** ________________________
- **Read-back, update and handover:** _______________________________
- **Welfare and learning:** ____

PRACTICE

A strong call record shows what was known, what remained uncertain, how risk was controlled and who explicitly accepted the next action.

FATIGUE CONCERN

Take self-report seriously, review task and relief arrangements and use the formal fitness route..

- **Establish exact location, immediate danger, source and caller needs.**
- **Choose the communication mode, authorised response and explicit owner.**
- **Read back critical detail, record action and confirm safe handover.**

CONTROL PRINCIPLE

Use role-specific scripts as prompts, not substitutes for listening. Any emergency, safeguarding or safety-critical action must follow current controlled procedure.

FATIGUE CONCERN - COMMUNICATION PRACTICE

Use this page for simulation, coaching or review. Do not enter real personal data in an uncontrolled workbook.

- **Caller, channel and exact location:** ______________________________
- **What is happening and immediate danger:** ________________________
- **Words that acknowledged and clarified:** __________________________
- **Operational action, authority and owner:** ________________________
- **Read-back, update and handover:** ________________________________
- **Welfare and learning:** ____

PRACTICE

A strong call record shows what was known, what remained uncertain, how risk was controlled and who explicitly accepted the next action.

VIOLENCE AND PERSONAL SAFETY

Support withdrawal, police or local response and medical care without expecting staff to remain exposed for service continuity..

- **Establish exact location, immediate danger, source and caller needs.**
- **Choose the communication mode, authorised response and explicit owner.**
- **Read back critical detail, record action and confirm safe handover.**

CONTROL PRINCIPLE

Use role-specific scripts as prompts, not substitutes for listening. Any emergency, safeguarding or safety-critical action must follow current controlled procedure.

VIOLENCE AND PERSONAL SAFETY - COMMUNICATION PRACTICE

Use this page for simulation, coaching or review. Do not enter real personal data in an uncontrolled workbook.

- **Caller, channel and exact location: ______________________________**
- **What is happening and immediate danger: _________________________**
- **Words that acknowledged and clarified: __________________________**
- **Operational action, authority and owner: _________________________**
- **Read-back, update and handover: ________________________________**
- **Welfare and learning: ____**

PRACTICE

A strong call record shows what was known, what remained uncertain, how risk was controlled and who explicitly accepted the next action.

WELFARE AFTER INCIDENT

Check fitness to continue, travel and support and arrange later follow-up rather than a single immediate question..

- **Establish exact location, immediate danger, source and caller needs.**
- **Choose the communication mode, authorised response and explicit owner.**
- **Read back critical detail, record action and confirm safe handover.**

CONTROL PRINCIPLE

Use role-specific scripts as prompts, not substitutes for listening. Any emergency, safeguarding or safety-critical action must follow current controlled procedure.

WELFARE AFTER INCIDENT - COMMUNICATION PRACTICE

Use this page for simulation, coaching or review. Do not enter real personal data in an uncontrolled workbook.

- **Caller, channel and exact location:** ______________________________
- **What is happening and immediate danger:** ________________________
- **Words that acknowledged and clarified:** __________________________
- **Operational action, authority and owner:** _______________________
- **Read-back, update and handover:** ________________________________
- **Welfare and learning:** ____

PRACTICE

A strong call record shows what was known, what remained uncertain, how risk was controlled and who explicitly accepted the next action.

PART 16 | INCIDENT COORDINATION

SHARED AWARENESS, CLEAR COMMAND, RECORDED DECISIONS

Remote coordination succeeds when each participant knows the situation, authority, action and next update.

STRUCTURED INITIAL REPORT

Capture location, event, hazards, people, operational effect, actions and support requested..

- **Establish exact location, immediate danger, source and caller needs.**
- **Choose the communication mode, authorised response and explicit owner.**
- **Read back critical detail, record action and confirm safe handover.**

CONTROL PRINCIPLE

Use role-specific scripts as prompts, not substitutes for listening. Any emergency, safeguarding or safety-critical action must follow current controlled procedure.

STRUCTURED INITIAL REPORT - COMMUNICATION PRACTICE

Use this page for simulation, coaching or review. Do not enter real personal data in an uncontrolled workbook.

- **Caller, channel and exact location:** ________________________________
- **What is happening and immediate danger:** ____________________________
- **Words that acknowledged and clarified:** _____________________________
- **Operational action, authority and owner:** ___________________________
- **Read-back, update and handover:** ___________________________________
- **Welfare and learning:** ____

PRACTICE

A strong call record shows what was known, what remained uncertain, how risk was controlled and who explicitly accepted the next action.

COMMAND AND CONTROL

Identify operational and emergency command roles and avoid issuing outside authority..

- **Establish exact location, immediate danger, source and caller needs.**
- **Choose the communication mode, authorised response and explicit owner.**
- **Read back critical detail, record action and confirm safe handover.**

CONTROL PRINCIPLE

Use role-specific scripts as prompts, not substitutes for listening. Any emergency, safeguarding or safety-critical action must follow current controlled procedure.

COMMAND AND CONTROL - COMMUNICATION PRACTICE

Use this page for simulation, coaching or review. Do not enter real personal data in an uncontrolled workbook.

- **Caller, channel and exact location:** ______________________________
- **What is happening and immediate danger:** ________________________
- **Words that acknowledged and clarified:** __________________________
- **Operational action, authority and owner:** ________________________
- **Read-back, update and handover:** _______________________________
- **Welfare and learning:** ____

PRACTICE

A strong call record shows what was known, what remained uncertain, how risk was controlled and who explicitly accepted the next action.

RESOURCE COORDINATION

Track who was requested, accepted, deployed, arrived and completed rather than treating notification as delivery..

- **Establish exact location, immediate danger, source and caller needs.**
- **Choose the communication mode, authorised response and explicit owner.**
- **Read back critical detail, record action and confirm safe handover.**

CONTROL PRINCIPLE

Use role-specific scripts as prompts, not substitutes for listening. Any emergency, safeguarding or safety-critical action must follow current controlled procedure.

RESOURCE COORDINATION - COMMUNICATION PRACTICE

Use this page for simulation, coaching or review. Do not enter real personal data in an uncontrolled workbook.

- **Caller, channel and exact location:** ______________________________
- **What is happening and immediate danger:** ________________________
- **Words that acknowledged and clarified:** __________________________
- **Operational action, authority and owner:** ________________________
- **Read-back, update and handover:** ________________________________
- **Welfare and learning:** ____

PRACTICE

A strong call record shows what was known, what remained uncertain, how risk was controlled and who explicitly accepted the next action.

DECISION LOG

Record time, information, decision, rationale, owner and review using approved systems..

- **Establish exact location, immediate danger, source and caller needs.**
- **Choose the communication mode, authorised response and explicit owner.**
- **Read back critical detail, record action and confirm safe handover.**

CONTROL PRINCIPLE

Use role-specific scripts as prompts, not substitutes for listening. Any emergency, safeguarding or safety-critical action must follow current controlled procedure.

DECISION LOG - COMMUNICATION PRACTICE

Use this page for simulation, coaching or review. Do not enter real personal data in an uncontrolled workbook.

- **Caller, channel and exact location:** ________________________________
- **What is happening and immediate danger:** __________________________
- **Words that acknowledged and clarified:** ____________________________
- **Operational action, authority and owner:** __________________________
- **Read-back, update and handover:** __________________________________
- **Welfare and learning:** ____

PRACTICE

A strong call record shows what was known, what remained uncertain, how risk was controlled and who explicitly accepted the next action.

REGULAR UPDATES

Use an agreed rhythm and send material change immediately without flooding recipients with repetition..

- **Establish exact location, immediate danger, source and caller needs.**
- **Choose the communication mode, authorised response and explicit owner.**
- **Read back critical detail, record action and confirm safe handover.**

CONTROL PRINCIPLE

Use role-specific scripts as prompts, not substitutes for listening. Any emergency, safeguarding or safety-critical action must follow current controlled procedure.

REGULAR UPDATES - COMMUNICATION PRACTICE

Use this page for simulation, coaching or review. Do not enter real personal data in an uncontrolled workbook.

- **Caller, channel and exact location:** ______________________________
- **What is happening and immediate danger:** ________________________
- **Words that acknowledged and clarified:** __________________________
- **Operational action, authority and owner:** ________________________
- **Read-back, update and handover:** _______________________________
- **Welfare and learning:** ____

> **PRACTICE**
>
> **A strong call record shows what was known, what remained uncertain, how risk was controlled and who explicitly accepted the next action.**

RECOVERY HANDOVER

Confirm safe state, evidence, welfare, assets and outstanding actions before closing control involvement..

- **Establish exact location, immediate danger, source and caller needs.**
- **Choose the communication mode, authorised response and explicit owner.**
- **Read back critical detail, record action and confirm safe handover.**

CONTROL PRINCIPLE

Use role-specific scripts as prompts, not substitutes for listening. Any emergency, safeguarding or safety-critical action must follow current controlled procedure.

RECOVERY HANDOVER - COMMUNICATION PRACTICE

Use this page for simulation, coaching or review. Do not enter real personal data in an uncontrolled workbook.

- **Caller, channel and exact location:** ________________________________
- **What is happening and immediate danger:** __________________________
- **Words that acknowledged and clarified:** ____________________________
- **Operational action, authority and owner:** __________________________
- **Read-back, update and handover:** ___________________________________
- **Welfare and learning:** ____

PRACTICE

A strong call record shows what was known, what remained uncertain, how risk was controlled and who explicitly accepted the next action.

PART 17 | TECHNOLOGY AND CONTINUITY

SYSTEMS SUPPORT CONTROL; THEY DO NOT OWN IT

Telephony, CCTV, radio and assistance systems can fail, lag or present conflicting information.

TELEPHONY AND RECORDING

Know recording status, emergency location limits, call transfer and fallback routes..

- **Establish exact location, immediate danger, source and caller needs.**
- **Choose the communication mode, authorised response and explicit owner.**
- **Read back critical detail, record action and confirm safe handover.**

CONTROL PRINCIPLE

Use role-specific scripts as prompts, not substitutes for listening. Any emergency, safeguarding or safety-critical action must follow current controlled procedure.

TELEPHONY AND RECORDING - COMMUNICATION PRACTICE

Use this page for simulation, coaching or review. Do not enter real personal data in an uncontrolled workbook.

- **Caller, channel and exact location: ______________________________**
- **What is happening and immediate danger: ________________________**
- **Words that acknowledged and clarified: __________________________**
- **Operational action, authority and owner: ________________________**
- **Read-back, update and handover: ________________________________**
- **Welfare and learning: ____**

PRACTICE

A strong call record shows what was known, what remained uncertain, how risk was controlled and who explicitly accepted the next action.

CCTV LIMITATIONS

Check camera identity, time, blind spots and image quality; absence on screen is not proof of absence..

- **Establish exact location, immediate danger, source and caller needs.**
- **Choose the communication mode, authorised response and explicit owner.**
- **Read back critical detail, record action and confirm safe handover.**

CONTROL PRINCIPLE

Use role-specific scripts as prompts, not substitutes for listening. Any emergency, safeguarding or safety-critical action must follow current controlled procedure.

CCTV LIMITATIONS - COMMUNICATION PRACTICE

Use this page for simulation, coaching or review. Do not enter real personal data in an uncontrolled workbook.

- **Caller, channel and exact location:** ______________________________
- **What is happening and immediate danger:** ________________________
- **Words that acknowledged and clarified:** __________________________
- **Operational action, authority and owner:** ________________________
- **Read-back, update and handover:** ______________________________
- **Welfare and learning:** ____

PRACTICE

A strong call record shows what was known, what remained uncertain, how risk was controlled and who explicitly accepted the next action.

PASSENGER ASSIST SYSTEMS

Maintain accurate status and audit trail while using positive human handover for critical assistance..

- **Establish exact location, immediate danger, source and caller needs.**
- **Choose the communication mode, authorised response and explicit owner.**
- **Read back critical detail, record action and confirm safe handover.**

CONTROL PRINCIPLE

Use role-specific scripts as prompts, not substitutes for listening. Any emergency, safeguarding or safety-critical action must follow current controlled procedure.

PASSENGER ASSIST SYSTEMS - COMMUNICATION PRACTICE

Use this page for simulation, coaching or review. Do not enter real personal data in an uncontrolled workbook.

- **Caller, channel and exact location:** ______________________________
- **What is happening and immediate danger:** ______________________________
- **Words that acknowledged and clarified:** ______________________________
- **Operational action, authority and owner:** ______________________________
- **Read-back, update and handover:** ______________________________
- **Welfare and learning:** ______________________________

PRACTICE

A strong call record shows what was known, what remained uncertain, how risk was controlled and who explicitly accepted the next action.

HELP-POINT FAILURE

Provide degraded reporting and passenger support, record fault and assess risk created by reduced contact..

- **Establish exact location, immediate danger, source and caller needs.**
- **Choose the communication mode, authorised response and explicit owner.**
- **Read back critical detail, record action and confirm safe handover.**

CONTROL PRINCIPLE

Use role-specific scripts as prompts, not substitutes for listening. Any emergency, safeguarding or safety-critical action must follow current controlled procedure.

HELP-POINT FAILURE - COMMUNICATION PRACTICE

Use this page for simulation, coaching or review. Do not enter real personal data in an uncontrolled workbook.

- **Caller, channel and exact location:** ________________________________
- **What is happening and immediate danger:** __________________________
- **Words that acknowledged and clarified:** ____________________________
- **Operational action, authority and owner:** __________________________
- **Read-back, update and handover:** __________________________________
- **Welfare and learning:** ____

PRACTICE

A strong call record shows what was known, what remained uncertain, how risk was controlled and who explicitly accepted the next action.

CYBER AND DATA OUTAGE

Use authorised manual fallbacks, protect information and reconcile records after restoration..

- **Establish exact location, immediate danger, source and caller needs.**
- **Choose the communication mode, authorised response and explicit owner.**
- **Read back critical detail, record action and confirm safe handover.**

CONTROL PRINCIPLE

Use role-specific scripts as prompts, not substitutes for listening. Any emergency, safeguarding or safety-critical action must follow current controlled procedure.

CYBER AND DATA OUTAGE - COMMUNICATION PRACTICE

Use this page for simulation, coaching or review. Do not enter real personal data in an uncontrolled workbook.

- **Caller, channel and exact location:** ______________________________
- **What is happening and immediate danger:** _________________________
- **Words that acknowledged and clarified:** __________________________
- **Operational action, authority and owner:** ________________________
- **Read-back, update and handover:** ________________________________
- **Welfare and learning:** ____

PRACTICE

A strong call record shows what was known, what remained uncertain, how risk was controlled and who explicitly accepted the next action.

BUSINESS CONTINUITY

Exercise loss of building, people, power, network, voice and systems with clear invocation and recovery authority..

- **Establish exact location, immediate danger, source and caller needs.**
- **Choose the communication mode, authorised response and explicit owner.**
- **Read back critical detail, record action and confirm safe handover.**

CONTROL PRINCIPLE

Use role-specific scripts as prompts, not substitutes for listening. Any emergency, safeguarding or safety-critical action must follow current controlled procedure.

BUSINESS CONTINUITY - COMMUNICATION PRACTICE

Use this page for simulation, coaching or review. Do not enter real personal data in an uncontrolled workbook.

- **Caller, channel and exact location:** ______________________________
- **What is happening and immediate danger:** ________________________
- **Words that acknowledged and clarified:** __________________________
- **Operational action, authority and owner:** ________________________
- **Read-back, update and handover:** ________________________________
- **Welfare and learning:** ____

> **PRACTICE**
>
> **A strong call record shows what was known, what remained uncertain, how risk was controlled and who explicitly accepted the next action.**

PART 18 | GDPR AND RECORDINGS

RECORD WHAT IS NECESSARY, PROTECT WHAT IS SENSITIVE

Control centres handle voice, health, disability, location, CCTV and incident information that can materially affect people.

TRANSPARENCY

Explain recording and data use through approved notices without delaying an immediate emergency response..

- **Establish exact location, immediate danger, source and caller needs.**
- **Choose the communication mode, authorised response and explicit owner.**
- **Read back critical detail, record action and confirm safe handover.**

CONTROL PRINCIPLE

Use role-specific scripts as prompts, not substitutes for listening. Any emergency, safeguarding or safety-critical action must follow current controlled procedure.

TRANSPARENCY - COMMUNICATION PRACTICE

Use this page for simulation, coaching or review. Do not enter real personal data in an uncontrolled workbook.

- **Caller, channel and exact location: ____________________________**
- **What is happening and immediate danger: _______________________**
- **Words that acknowledged and clarified: _________________________**
- **Operational action, authority and owner: _______________________**
- **Read-back, update and handover: ______________________________**
- **Welfare and learning: _______________________________________**

PRACTICE

A strong call record shows what was known, what remained uncertain, how risk was controlled and who explicitly accepted the next action.

DATA MINIMISATION

Capture what is necessary for assistance, safety, evidence and legitimate follow-up - not curiosity..

- **Establish exact location, immediate danger, source and caller needs.**
- **Choose the communication mode, authorised response and explicit owner.**
- **Read back critical detail, record action and confirm safe handover.**

CONTROL PRINCIPLE

Use role-specific scripts as prompts, not substitutes for listening. Any emergency, safeguarding or safety-critical action must follow current controlled procedure.

DATA MINIMISATION - COMMUNICATION PRACTICE

Use this page for simulation, coaching or review. Do not enter real personal data in an uncontrolled workbook.

- **Caller, channel and exact location:** ______________________________
- **What is happening and immediate danger:** ___________________________
- **Words that acknowledged and clarified:** ____________________________
- **Operational action, authority and owner:** ___________________________
- **Read-back, update and handover:** _________________________________
- **Welfare and learning:** ____

PRACTICE

A strong call record shows what was known, what remained uncertain, how risk was controlled and who explicitly accepted the next action.

HEALTH AND DISABILITY DATA

Use additional safeguards, restricted access and purpose-limited sharing for special-category information..

- **Establish exact location, immediate danger, source and caller needs.**
- **Choose the communication mode, authorised response and explicit owner.**
- **Read back critical detail, record action and confirm safe handover.**

CONTROL PRINCIPLE

Use role-specific scripts as prompts, not substitutes for listening. Any emergency, safeguarding or safety-critical action must follow current controlled procedure.

HEALTH AND DISABILITY DATA - COMMUNICATION PRACTICE

Use this page for simulation, coaching or review. Do not enter real personal data in an uncontrolled workbook.

- **Caller, channel and exact location: ______________________________**
- **What is happening and immediate danger: ______________________________**
- **Words that acknowledged and clarified: ______________________________**
- **Operational action, authority and owner: ______________________________**
- **Read-back, update and handover: ______________________________**
- **Welfare and learning: ______________________________**

PRACTICE

A strong call record shows what was known, what remained uncertain, how risk was controlled and who explicitly accepted the next action.

POLICE AND THIRD-PARTY REQUESTS

Use authorised disclosure routes and record what was shared, why and by whom..

- **Establish exact location, immediate danger, source and caller needs.**
- **Choose the communication mode, authorised response and explicit owner.**
- **Read back critical detail, record action and confirm safe handover.**

CONTROL PRINCIPLE

Use role-specific scripts as prompts, not substitutes for listening. Any emergency, safeguarding or safety-critical action must follow current controlled procedure.

POLICE AND THIRD-PARTY REQUESTS - COMMUNICATION PRACTICE

Use this page for simulation, coaching or review. Do not enter real personal data in an uncontrolled workbook.

- **Caller, channel and exact location: ______________________**
- **What is happening and immediate danger: __________________**
- **Words that acknowledged and clarified: ____________________**
- **Operational action, authority and owner: __________________**
- **Read-back, update and handover: _________________________**
- **Welfare and learning: ______________________________**

PRACTICE

A strong call record shows what was known, what remained uncertain, how risk was controlled and who explicitly accepted the next action.

SUBJECT ACCESS AND RETENTION

Preserve searchable approved records and follow organisational retention and rights processes..

- **Establish exact location, immediate danger, source and caller needs.**
- **Choose the communication mode, authorised response and explicit owner.**
- **Read back critical detail, record action and confirm safe handover.**

CONTROL PRINCIPLE

Use role-specific scripts as prompts, not substitutes for listening. Any emergency, safeguarding or safety-critical action must follow current controlled procedure.

SUBJECT ACCESS AND RETENTION - COMMUNICATION PRACTICE

Use this page for simulation, coaching or review. Do not enter real personal data in an uncontrolled workbook.

- **Caller, channel and exact location:** ______________________________
- **What is happening and immediate danger:** ______________________________
- **Words that acknowledged and clarified:** ______________________________
- **Operational action, authority and owner:** ______________________________
- **Read-back, update and handover:** ______________________________
- **Welfare and learning:** ______________________________

PRACTICE

A strong call record shows what was known, what remained uncertain, how risk was controlled and who explicitly accepted the next action.

DATA BREACH

Contain wrong disclosure, preserve facts and notify the information team promptly rather than quietly deleting evidence..

- **Establish exact location, immediate danger, source and caller needs.**
- **Choose the communication mode, authorised response and explicit owner.**
- **Read back critical detail, record action and confirm safe handover.**

CONTROL PRINCIPLE

Use role-specific scripts as prompts, not substitutes for listening. Any emergency, safeguarding or safety-critical action must follow current controlled procedure.

DATA BREACH - COMMUNICATION PRACTICE

Use this page for simulation, coaching or review. Do not enter real personal data in an uncontrolled workbook.

- **Caller, channel and exact location:** ______________________________
- **What is happening and immediate danger:** ________________________
- **Words that acknowledged and clarified:** __________________________
- **Operational action, authority and owner:** ________________________
- **Read-back, update and handover:** ________________________________
- **Welfare and learning:** ____

PRACTICE

A strong call record shows what was known, what remained uncertain, how risk was controlled and who explicitly accepted the next action.

THE CONTROLLER IS PART OF THE SAFETY SYSTEM

Attention, workload, fatigue and emotional exposure affect remote judgement even when procedures are correct.

WORKLOAD AND TASK SWITCHING

Control queue, priorities, interruptions and handoffs rather than expecting memory to carry multiple evolving incidents..

- **Establish exact location, immediate danger, source and caller needs.**
- **Choose the communication mode, authorised response and explicit owner.**
- **Read back critical detail, record action and confirm safe handover.**

CONTROL PRINCIPLE

Use role-specific scripts as prompts, not substitutes for listening. Any emergency, safeguarding or safety-critical action must follow current controlled procedure.

WORKLOAD AND TASK SWITCHING - COMMUNICATION PRACTICE

Use this page for simulation, coaching or review. Do not enter real personal data in an uncontrolled workbook.

- **Caller, channel and exact location:** ______________________________
- **What is happening and immediate danger:** ________________________
- **Words that acknowledged and clarified:** __________________________
- **Operational action, authority and owner:** ________________________
- **Read-back, update and handover:** _______________________________
- **Welfare and learning:** ____

PRACTICE

A strong call record shows what was known, what remained uncertain, how risk was controlled and who explicitly accepted the next action.

ALARM AND SCREEN LOAD

Design and supervise work so critical cues are distinguishable and operators can maintain awareness..

- **Establish exact location, immediate danger, source and caller needs.**
- **Choose the communication mode, authorised response and explicit owner.**
- **Read back critical detail, record action and confirm safe handover.**

CONTROL PRINCIPLE

Use role-specific scripts as prompts, not substitutes for listening. Any emergency, safeguarding or safety-critical action must follow current controlled procedure.

ALARM AND SCREEN LOAD - COMMUNICATION PRACTICE

Use this page for simulation, coaching or review. Do not enter real personal data in an uncontrolled workbook.

- **Caller, channel and exact location:** ______________________________
- **What is happening and immediate danger:** ________________________
- **Words that acknowledged and clarified:** __________________________
- **Operational action, authority and owner:** ________________________
- **Read-back, update and handover:** ________________________________
- **Welfare and learning:** ____

PRACTICE

A strong call record shows what was known, what remained uncertain, how risk was controlled and who explicitly accepted the next action.

SHIFT AND FATIGUE

Plan nights, breaks, overtime and recovery through the fatigue-risk system; personal resilience is not a substitute..

- **Establish exact location, immediate danger, source and caller needs.**
- **Choose the communication mode, authorised response and explicit owner.**
- **Read back critical detail, record action and confirm safe handover.**

CONTROL PRINCIPLE

Use role-specific scripts as prompts, not substitutes for listening. Any emergency, safeguarding or safety-critical action must follow current controlled procedure.

SHIFT AND FATIGUE - COMMUNICATION PRACTICE

Use this page for simulation, coaching or review. Do not enter real personal data in an uncontrolled workbook.

- **Caller, channel and exact location:** ______________________________
- **What is happening and immediate danger:** ______________________________
- **Words that acknowledged and clarified:** ______________________________
- **Operational action, authority and owner:** ______________________________
- **Read-back, update and handover:** ______________________________
- **Welfare and learning:** ______________________________

PRACTICE

A strong call record shows what was known, what remained uncertain, how risk was controlled and who explicitly accepted the next action.

DISTRESSING CALLS

Provide immediate supervisor contact, decompression and a route to trauma or occupational-health support..

- **Establish exact location, immediate danger, source and caller needs.**
- **Choose the communication mode, authorised response and explicit owner.**
- **Read back critical detail, record action and confirm safe handover.**

> **CONTROL PRINCIPLE**
>
> **Use role-specific scripts as prompts, not substitutes for listening. Any emergency, safeguarding or safety-critical action must follow current controlled procedure.**

DISTRESSING CALLS - COMMUNICATION PRACTICE

Use this page for simulation, coaching or review. Do not enter real personal data in an uncontrolled workbook.

- **Caller, channel and exact location:** ______________________________
- **What is happening and immediate danger:** ________________________
- **Words that acknowledged and clarified:** __________________________
- **Operational action, authority and owner:** ________________________
- **Read-back, update and handover:** ________________________________
- **Welfare and learning:** ____

PRACTICE

A strong call record shows what was known, what remained uncertain, how risk was controlled and who explicitly accepted the next action.

FITNESS TO CONTINUE

Reallocate or relieve a controller whose attention, distress or fatigue may affect safe performance..

- **Establish exact location, immediate danger, source and caller needs.**
- **Choose the communication mode, authorised response and explicit owner.**
- **Read back critical detail, record action and confirm safe handover.**

CONTROL PRINCIPLE

Use role-specific scripts as prompts, not substitutes for listening. Any emergency, safeguarding or safety-critical action must follow current controlled procedure.

FITNESS TO CONTINUE - COMMUNICATION PRACTICE

Use this page for simulation, coaching or review. Do not enter real personal data in an uncontrolled workbook.

- **Caller, channel and exact location:** ______________________________
- **What is happening and immediate danger:** ______________________________
- **Words that acknowledged and clarified:** ______________________________
- **Operational action, authority and owner:** ______________________________
- **Read-back, update and handover:** ______________________________
- **Welfare and learning:** ______________________________

PRACTICE

A strong call record shows what was known, what remained uncertain, how risk was controlled and who explicitly accepted the next action.

TEAM WELFARE CULTURE

Normalise speaking up, peer checking and fair support without forcing staff to disclose private clinical detail..

- **Establish exact location, immediate danger, source and caller needs.**
- **Choose the communication mode, authorised response and explicit owner.**
- **Read back critical detail, record action and confirm safe handover.**

CONTROL PRINCIPLE

Use role-specific scripts as prompts, not substitutes for listening. Any emergency, safeguarding or safety-critical action must follow current controlled procedure.

TEAM WELFARE CULTURE - COMMUNICATION PRACTICE

Use this page for simulation, coaching or review. Do not enter real personal data in an uncontrolled workbook.

- **Caller, channel and exact location: ______________________________**
- **What is happening and immediate danger: ________________________**
- **Words that acknowledged and clarified: __________________________**
- **Operational action, authority and owner: ________________________**
- **Read-back, update and handover: ________________________________**
- **Welfare and learning: ____**

PRACTICE

A strong call record shows what was known, what remained uncertain, how risk was controlled and who explicitly accepted the next action.

PART 20 | COMPETENCE AND ASSURANCE

PRACTISE THE DIFFICULT CALL BEFORE IT ARRIVES

Competence requires knowledge, simulated performance, observation, feedback, reassessment and learning from real work.

ROLE COMPETENCE

Define technical, communication, accessibility, system, emergency and welfare capabilities for each control function..

- **Establish exact location, immediate danger, source and caller needs.**
- **Choose the communication mode, authorised response and explicit owner.**
- **Read back critical detail, record action and confirm safe handover.**

CONTROL PRINCIPLE

Use role-specific scripts as prompts, not substitutes for listening. Any emergency, safeguarding or safety-critical action must follow current controlled procedure.

ROLE COMPETENCE - COMMUNICATION PRACTICE

Use this page for simulation, coaching or review. Do not enter real personal data in an uncontrolled workbook.

- **Caller, channel and exact location: ________________________________**
- **What is happening and immediate danger: ____________________________**
- **Words that acknowledged and clarified: ______________________________**
- **Operational action, authority and owner: ___________________________**
- **Read-back, update and handover: ____________________________________**
- **Welfare and learning: ____**

PRACTICE

A strong call record shows what was known, what remained uncertain, how risk was controlled and who explicitly accepted the next action.

SIMULATION

Practise uncertainty, silence, wrong location, distressed callers, system failure and multi-agency handover..

- **Establish exact location, immediate danger, source and caller needs.**
- **Choose the communication mode, authorised response and explicit owner.**
- **Read back critical detail, record action and confirm safe handover.**

CONTROL PRINCIPLE

Use role-specific scripts as prompts, not substitutes for listening. Any emergency, safeguarding or safety-critical action must follow current controlled procedure.

SIMULATION - COMMUNICATION PRACTICE

Use this page for simulation, coaching or review. Do not enter real personal data in an uncontrolled workbook.

- **Caller, channel and exact location: ________________________________**
- **What is happening and immediate danger: __________________________**
- **Words that acknowledged and clarified: ____________________________**
- **Operational action, authority and owner: __________________________**
- **Read-back, update and handover: _________________________________**
- **Welfare and learning: ____**

PRACTICE

A strong call record shows what was known, what remained uncertain, how risk was controlled and who explicitly accepted the next action.

CALL REVIEW

Sample proportionately, protect privacy and assess listening, risk, accuracy, ownership and outcome..

- **Establish exact location, immediate danger, source and caller needs.**
- **Choose the communication mode, authorised response and explicit owner.**
- **Read back critical detail, record action and confirm safe handover.**

CONTROL PRINCIPLE

Use role-specific scripts as prompts, not substitutes for listening. Any emergency, safeguarding or safety-critical action must follow current controlled procedure.

CALL REVIEW - COMMUNICATION PRACTICE

Use this page for simulation, coaching or review. Do not enter real personal data in an uncontrolled workbook.

- **Caller, channel and exact location:** ______________________________
- **What is happening and immediate danger:** ______________________________
- **Words that acknowledged and clarified:** ______________________________
- **Operational action, authority and owner:** ______________________________
- **Read-back, update and handover:** ______________________________
- **Welfare and learning:** ______________________________

PRACTICE

A strong call record shows what was known, what remained uncertain, how risk was controlled and who explicitly accepted the next action.

COACHING

Use specific moments and alternatives; do not reduce human communication to script compliance..

- **Establish exact location, immediate danger, source and caller needs.**
- **Choose the communication mode, authorised response and explicit owner.**
- **Read back critical detail, record action and confirm safe handover.**

CONTROL PRINCIPLE

Use role-specific scripts as prompts, not substitutes for listening. Any emergency, safeguarding or safety-critical action must follow current controlled procedure.

COACHING - COMMUNICATION PRACTICE

Use this page for simulation, coaching or review. Do not enter real personal data in an uncontrolled workbook.

- **Caller, channel and exact location:** ______________________________
- **What is happening and immediate danger:** ______________________________
- **Words that acknowledged and clarified:** ______________________________
- **Operational action, authority and owner:** ______________________________
- **Read-back, update and handover:** ______________________________
- **Welfare and learning:** ______________________________

PRACTICE

A strong call record shows what was known, what remained uncertain, how risk was controlled and who explicitly accepted the next action.

FAIR ACCOUNTABILITY

Distinguish system design, workload, training, error, adaptation and deliberate breach using evidence..

- **Establish exact location, immediate danger, source and caller needs.**
- **Choose the communication mode, authorised response and explicit owner.**
- **Read back critical detail, record action and confirm safe handover.**

CONTROL PRINCIPLE

Use role-specific scripts as prompts, not substitutes for listening. Any emergency, safeguarding or safety-critical action must follow current controlled procedure.

FAIR ACCOUNTABILITY - COMMUNICATION PRACTICE

Use this page for simulation, coaching or review. Do not enter real personal data in an uncontrolled workbook.

- **Caller, channel and exact location: ______________________________**
- **What is happening and immediate danger: ________________________**
- **Words that acknowledged and clarified: __________________________**
- **Operational action, authority and owner: ________________________**
- **Read-back, update and handover: _______________________________**
- **Welfare and learning: ____**

PRACTICE

A strong call record shows what was known, what remained uncertain, how risk was controlled and who explicitly accepted the next action.

CONTINUOUS IMPROVEMENT

Track failed handovers, complaints, near misses and welfare data to verify whether changes reduce risk..

- **Establish exact location, immediate danger, source and caller needs.**
- **Choose the communication mode, authorised response and explicit owner.**
- **Read back critical detail, record action and confirm safe handover.**

CONTROL PRINCIPLE

Use role-specific scripts as prompts, not substitutes for listening. Any emergency, safeguarding or safety-critical action must follow current controlled procedure.

CONTINUOUS IMPROVEMENT - COMMUNICATION PRACTICE

Use this page for simulation, coaching or review. Do not enter real personal data in an uncontrolled workbook.

- **Caller, channel and exact location: ____________________________________**
- **What is happening and immediate danger: ______________________________**
- **Words that acknowledged and clarified: ________________________________**
- **Operational action, authority and owner: ______________________________**
- **Read-back, update and handover: ______________________________________**
- **Welfare and learning: ____**

PRACTICE

A strong call record shows what was known, what remained uncertain, how risk was controlled and who explicitly accepted the next action.

CONTROL CARDS

QUICK-REFERENCE PLAYBOOKS

Compact prompts for training and safe recall. Local controlled procedures remain authoritative.

FIRST-CONTACT CARD

Use the card to prepare or review communication. Do not read it mechanically when the caller provides more urgent information.

- **Identify yourself and function.**
- **Exact location and landmark.**
- **Immediate danger or injury.**
- **What is happening now.**
- **Callback or fallback contact.**
- **Action, owner and next update.**

QUICK CHECK

No card grants authority or replaces competence. Confirm current local escalation, recording and handover requirements.

REMOTE LOCATION CHECK

Use the card to prepare or review communication. Do not read it mechanically when the caller provides more urgent information.

- **Station and entrance.**
- **Platform and direction.**
- **Train, coach and door.**
- **Sign, help-point or asset number.**
- **What the caller can see or hear.**
- **Location repeated back and confirmed.**

QUICK CHECK

No card grants authority or replaces competence. Confirm current local escalation, recording and handover requirements.

DISTRESS CONVERSATION

Use the card to prepare or review communication. Do not read it mechanically when the caller provides more urgent information.

- **Slow voice and one question at a time.**
- **Acknowledge impact without false reassurance.**
- **Orient to location and present danger.**
- **Offer only a known-safe next step.**
- **Explain silence and action.**
- **Maintain connection through handover.**

QUICK CHECK

No card grants authority or replaces competence. Confirm current local escalation, recording and handover requirements.

SUICIDE CONCERN

Use the card to prepare or review communication. Do not read it mechanically when the caller provides more urgent information.

- **Ask directly about suicide or self-harm.**
- **Exact current location and immediate intention.**
- **Dangerous access and whether alone.**
- **Keep listening without judgement.**
- **Activate trained emergency procedure.**
- **Do not promise secrecy; confirm handover.**

QUICK CHECK

No card grants authority or replaces competence. Confirm current local escalation, recording and handover requirements.

SAFETY-CRITICAL MESSAGE

Use the card to prepare or review communication. Do not read it mechanically when the caller provides more urgent information.

- **Correct identity and authority.**
- **Exact location, train or asset.**
- **Accurate, brief and clear content.**
- **Required action and limit.**
- **Read-back and hear-back.**
- **Time and record.**

QUICK CHECK

No card grants authority or replaces competence. Confirm current local escalation, recording and handover requirements.

PASSENGER ASSIST HANDOVER

Use the card to prepare or review communication. Do not read it mechanically when the caller provides more urgent information.

- **Passenger identifier and contact.**
- **Service, arrival and destination.**
- **Coach, door and meeting point.**
- **Assistance and equipment.**
- **Named or functional acceptance.**
- **Contingency and completion.**

QUICK CHECK

No card grants authority or replaces competence. Confirm current local escalation, recording and handover requirements.

ASSISTANCE FAILURE

Use the card to prepare or review communication. Do not read it mechanically when the caller provides more urgent information.

- **Where is the passenger now?**
- **Are they safe and able to disembark?**
- **Immediate accessible support.**
- **Onward, medication or care consequence.**
- **Owner and timed update.**
- **Failure record and review.**

QUICK CHECK

No card grants authority or replaces competence. Confirm current local escalation, recording and handover requirements.

SAFEGUARDING REPORT

Use the card to prepare or review communication. Do not read it mechanically when the caller provides more urgent information.

- **Exact person and location.**
- **Immediate safety and supervision.**
- **Observed behaviour or disclosure.**
- **Associates and direction.**
- **Information shared and authority.**
- **Handover and reference.**

QUICK CHECK

No card grants authority or replaces competence. Confirm current local escalation, recording and handover requirements.

AGGRESSIVE CALLER

Use the card to prepare or review communication. Do not read it mechanically when the caller provides more urgent information.

- **Acknowledge frustration.**
- **Identify actionable problem.**
- **Set behavioural boundary.**
- **Offer realistic choices.**
- **Escalate credible threat.**
- **Record exact words and aftercare.**

QUICK CHECK

No card grants authority or replaces competence. Confirm current local escalation, recording and handover requirements.

STAFF UNCERTAINTY

Use the card to prepare or review communication. Do not read it mechanically when the caller provides more urgent information.

- **Location and safe state.**
- **Immediate hazard.**
- **What is known and uncertain.**
- **Authority or competence required.**
- **Do not continue until confirmed.**
- **Welfare and follow-up.**

QUICK CHECK

No card grants authority or replaces competence. Confirm current local escalation, recording and handover requirements.

INCIDENT UPDATE

Use the card to prepare or review communication. Do not read it mechanically when the caller provides more urgent information.

- **Location and current situation.**
- **People and service effect.**
- **Hazards and controls.**
- **Resources requested and arrived.**
- **Command and decisions required.**
- **Next update time.**

QUICK CHECK

No card grants authority or replaces competence. Confirm current local escalation, recording and handover requirements.

SYSTEM FAILURE

Use the card to prepare or review communication. Do not read it mechanically when the caller provides more urgent information.

- **Failed function and time.**
- **Operational and passenger consequence.**
- **Fallback invoked.**
- **Information reconciliation.**
- **Technical owner and escalation.**
- **Restoration test and closure.**

QUICK CHECK

No card grants authority or replaces competence. Confirm current local escalation, recording and handover requirements.

CONTROLLER WELFARE

Use the card to prepare or review communication. Do not read it mechanically when the caller provides more urgent information.

- **Nature and duration of exposure.**
- **Immediate physical or emotional need.**
- **Fitness to continue.**
- **Relief and decompression.**
- **Peer, manager or professional support.**
- **Confidential follow-up.**

QUICK CHECK

No card grants authority or replaces competence. Confirm current local escalation, recording and handover requirements.

SHIFT HANDOVER

Use the card to prepare or review communication. Do not read it mechanically when the caller provides more urgent information.

- **Open calls and incidents.**
- **People awaiting assistance or update.**
- **System failures and workarounds.**
- **Emergency or safeguarding referrals.**
- **Outstanding owner and deadline.**
- **Controller welfare concerns.**

QUICK CHECK

No card grants authority or replaces competence. Confirm current local escalation, recording and handover requirements.

AVOID THESE PHRASES

Use the card to prepare or review communication. Do not read it mechanically when the caller provides more urgent information.

- **Calm down.**
- **Do not worry.**
- **You will be fine.**
- **I know exactly how you feel.**
- **There is nothing we can do.**
- **At least nobody was hurt.**

QUICK CHECK

No card grants authority or replaces competence. Confirm current local escalation, recording and handover requirements.

HARVARD REFERENCE LIST

Authoritative starting points for rail communication, assistance, distress support, competence, fatigue and information governance.

RAIL SAFETY AND STANDARDS BOARD

Safety-critical communications learning material and current standards catalogue. London: RSSB, accessed 13 August 2026.

- **Confirm current issue and effective date.**
- **Check role, operator and location applicability.**
- **Use the organisation's controlled implementation.**

SOURCE NOTE

This independent guide is not affiliated with or endorsed by the cited organisations.

OFFICE OF RAIL AND ROAD

Accessible Travel Policy: Guidance for Train and Station Operators. October 2025 edition. London: ORR.

- **Confirm current issue and effective date.**
- **Check role, operator and location applicability.**
- **Use the organisation's controlled implementation.**

SOURCE NOTE

This independent guide is not affiliated with or endorsed by the cited organisations.

OFFICE OF RAIL AND ROAD

Communications between stations when assisting passengers and the Passenger Assist Handover Protocol. London: ORR, accessed 13 August 2026.

- **Confirm current issue and effective date.**
- **Check role, operator and location applicability.**
- **Use the organisation's controlled implementation.**

SOURCE NOTE

This independent guide is not affiliated with or endorsed by the cited organisations.

OFFICE OF RAIL AND ROAD

Managing Rail Staff Fatigue. London: ORR, current online edition.

- **Confirm current issue and effective date.**
- **Check role, operator and location applicability.**
- **Use the organisation's controlled implementation.**

SOURCE NOTE

This independent guide is not affiliated with or endorsed by the cited organisations.

OFFICE OF RAIL AND ROAD

Developing and Maintaining Staff Competence. Railway Safety Publication 1. London: ORR.

- **Confirm current issue and effective date.**
- **Check role, operator and location applicability.**
- **Use the organisation's controlled implementation.**

SOURCE NOTE

This independent guide is not affiliated with or endorsed by the cited organisations.

SAMARITANS

Supporting someone with suicidal thoughts; listening without judgement; and asking directly about suicide. Available at: https://www.samaritans.org/ (Accessed: 13 August 2026).

- **Confirm current issue and effective date.**
- **Check role, operator and location applicability.**
- **Use the organisation's controlled implementation.**

SOURCE NOTE

This independent guide is not affiliated with or endorsed by the cited organisations.

RAIL DELIVERY GROUP

Passenger Assist: shared industry assistance system and passenger information. London: RDG, current material.

- **Confirm current issue and effective date.**
- **Check role, operator and location applicability.**
- **Use the organisation's controlled implementation.**

SOURCE NOTE

This independent guide is not affiliated with or endorsed by the cited organisations.

BRITISH TRANSPORT POLICE

Emergency and non-emergency reporting information, including text 61016. London: BTP, accessed 13 August 2026.

- **Confirm current issue and effective date.**
- **Check role, operator and location applicability.**
- **Use the organisation's controlled implementation.**

SOURCE NOTE

This independent guide is not affiliated with or endorsed by the cited organisations.

INFORMATION COMMISSIONER'S OFFICE

Employment, call recording, CCTV, special-category information and data-protection guidance. Wilmslow: ICO, accessed 13 August 2026.

- **Confirm current issue and effective date.**
- **Check role, operator and location applicability.**
- **Use the organisation's controlled implementation.**

SOURCE NOTE

This independent guide is not affiliated with or endorsed by the cited organisations.

UK PARLIAMENT

Railways and Other Guided Transport Systems (Safety) Regulations 2006, SI 2006/599, as amended.

- **Confirm current issue and effective date.**
- **Check role, operator and location applicability.**
- **Use the organisation's controlled implementation.**

SOURCE NOTE

This independent guide is not affiliated with or endorsed by the cited organisations.

UK PARLIAMENT

Data Protection Act 2018 and UK General Data Protection Regulation.

- **Confirm current issue and effective date.**
- **Check role, operator and location applicability.**
- **Use the organisation's controlled implementation.**

SOURCE NOTE

This independent guide is not affiliated with or endorsed by the cited organisations.

UK PARLIAMENT

Equality Act 2010. London: The Stationery Office.

- **Confirm current issue and effective date.**
- **Check role, operator and location applicability.**
- **Use the organisation's controlled implementation.**

SOURCE NOTE

This independent guide is not affiliated with or endorsed by the cited organisations.

HEALTH AND SAFETY EXECUTIVE

Work-related stress, shift work and first-aid needs guidance. Bootle: HSE, current online material.

- **Confirm current issue and effective date.**
- **Check role, operator and location applicability.**
- **Use the organisation's controlled implementation.**

SOURCE NOTE

This independent guide is not affiliated with or endorsed by the cited organisations.

THE CONTROLLER'S PROMISE

Remote communication becomes safe action when the controller hears the person, confirms the railway facts and keeps ownership visible until handover.

- **I will listen without losing operational focus.**
- **I will locate, confirm and read back.**
- **I will avoid false reassurance and unsupported promises.**
- **I will protect the caller, staff, information and myself.**

FINAL REFLECTION

Hear the person. Confirm the risk. Coordinate the safe response.

ABOUT THE AUTHOR

ROBERT PORTER

A senior operational leader with almost two decades across UK rail, transport and safety-critical environments. Robert has led frontline, station, incident and multi-site operations requiring calm remote coordination and clear communication.

ABOUT THIS PLAYBOOK

A practical guide to remote situational awareness, safety-critical communication, Passenger Assist, distress conversations, safeguarding, incident coordination and controller welfare.

THE PRACTICAL SAFETY SERIES

Hear the person. Confirm the risk.

Coordinate the safe response.

www.ingramcontent.com/pod-product-compliance
Lightning Source LLC
LaVergne TN
LVHW050804061126
851857LV00012B/675